Polymeric Drugs and Drug Delivery Systems

Polymeric Drugs and Drug Delivery Systems

Richard L. Dunn, EDITOR
Atrix Laboratories, Inc.

Raphael M. Ottenbrite, EDITOR
Virginia Commonwealth University

Developed from a symposium sponsored
by the Division of Polymer Chemistry, Inc.,
at the 200th National Meeting
of the American Chemical Society,
Washington, D.C.,
August 26–31, 1990

American Chemical Society, Washington, DC 1991

CHEM
Sep/ae

Library of Congress Cataloging-in-Publication Data

American Chemical Society. Meeting. (200th: 1990: Washington, D.C.)

Polymeric drugs and drug delivery systems / Richard L. Dunn, editor, Raphael M. Ottenbrite, editor; developed from a symposium sponsored by the Division of Polymeric Chemistry, Inc. at the 200th National Meeting of the American Chemical Society, Washington, D.C., August 26–31, 1990.

  p.   cm.—(ACS symposium series, ISSN 0097–6156; 469)

Includes bibliographical references and indexes.

ISBN 0–8412–2105–7

1. Polymeric drugs—Congresses. 2. Polymeric drug delivery systems—Congresses.

I. Dunn, Richard L. II. Ottenbrite, Raphael M. III. American Chemical Society. Division of Polymer Chemistry. IV. Title. V. Series.

RS201.P65A44   1991
615'.1—dc20

91–23423
CIP

SD 9/13/91 PL

ACS Symposium Series

M. Joan Comstock, *Series Editor*

1991 ACS Books Advisory Board

Foreword

THE ACS SYMPOSIUM SERIES was founded in 1974 to provide a medium for publishing symposia quickly in book form. The format of the Series parallels that of the continuing ADVANCES IN CHEMISTRY SERIES except that, in order to save time, the papers are not typeset, but are reproduced as they are submitted by the authors in camera-ready form. Papers are reviewed under the supervision of the editors with the assistance of the Advisory Board and are selected to maintain the integrity of the symposia. Both reviews and reports of research are acceptable, because symposia may embrace both types of presentation. However, verbatim reproductions of previously published papers are not accepted.

Contents

LIPOSOMAL DRUG DELIVERY

INDEXES

Preface

DELIVERY OF DRUGS by means of controlled-release technology began in the 1970s and has continued to expand so rapidly that there are now numerous products both on the market and in development. These controlled-release drug-delivery products have given new life to old pharmaceuticals that either were no longer patented or had properties that prevented them from being used effectively to treat various diseases. In addition to stimulating use of these older drug products, controlled-release technology is now being directed toward the newer biopharmaceuticals produced by genetic research. It is with biopharmaceuticals that controlled-release technology may find its most important applications in medicine.

Polymers have played a major role in the development of controlled-release systems and, as expected, the earlier polymeric drug-delivery systems incorporated polymers that were commercially available and approved by the U.S. Food and Drug Administration. There are many polymers that meet this need, and they have been successfully incorporated into commercial products for oral, injectable, implantable, topical, and transdermal administration. The mechanisms by which drugs are released from these polymers and the processes for fabricating such controlled-drug-delivery devices have been well reviewed in the literature. Extensive research efforts are being made to improve both the polymers and the processes, as well as to apply them to the controlled release of a wide variety of pharmaceutical products. However, with the continued development of controlled-release technology, the need has arisen for materials with more specific drug-delivery properties. These materials include new biodegradable polymers, polymers with both hydrophilic and hydrophobic characteristics, and hydrogels that respond to temperature or pH changes. In addition, methods to overcome some of the barriers associated with current drug-delivery systems are necessary. Finally, polymers that may not only be used to deliver drugs, but that may themselves elicit biological responses are needed.

This book is divided into four sections that cover the main topics in the field of drug delivery. The first section gives an overview of the polymers and materials currently being used in drug delivery and some of the problems with and opportunities for polymeric drug delivery. The overview chapters are followed by a section on polymeric drugs and polymer–drug conjugates. This section describes novel polymers that function as drugs themselves or that are covalently attached to drugs. This field of research offers tremendous possibilities for new materials that can be made either synthetically or by genetic engineering. The third section of the book deals with new polymers that can be used as matrices for drug delivery. Polymers described in this section are not covalently bound to a drug but rather are physically mixed or blended with the drug. Polymer–drug mixtures are currently the most widely used drug-delivery systems, and the chapters included here describe new materials that may be useful in the future. The final section covers new developments in the area of drug delivery with liposomes. This area has intrigued researchers for years, and with the development of new materials to target the liposomes, this field of research should remain prominent for the next several years.

As editors, we hope that this book will alert researchers to the problems associated with drug delivery and the opportunities for future developments. If the material presented here can stimulate new ideas and concepts for polymeric drugs and polymeric drug-delivery systems, then our efforts will have been worthwhile. We wish to thank the Division of Polymer Science, Inc., for sponsoring the symposium that served as the basis for this book, and Glaxo, Inc., Lilly Research Laboratories, and Atrix Laboratories, Inc., for providing partial funding. We also want to thank each of the authors for their participation and cooperation. Without them, this book would not have been possible. We gratefully acknowledge the staff of Atrix Laboratories and, specifically, Karen Miller and Sisca Wolff, for their assistance in the editing and production aspects of the book. Most of all, we want to thank Carol Dunn for her efforts in assembling and formatting all of the chapters and for her support throughout this endeavor.

RICHARD L. DUNN
Atrix Laboratories, Inc.
2579 Midpoint Drive
Fort Collins, CO 80525

RAPHAEL M. OTTENBRITE
Virginia Commonwealth University
Richmond, VA 23284

March 15, 1991

DRUG DELIVERY SYSTEMS

Drug Delivery Systems

The controlled delivery of drugs is a technology that draws on expertise from a number of scientific disciplines including chemistry, bioengineering, pharmacology, biology, polymer science and medicine. As the name implies, the objectives of controlled drug delivery are to disseminate a drug when and where it is needed and at the proper dose. Although mechanical devices such as pumps, syringes, and catheters can be used to deliver drugs, this book will focus on polymeric and liposomal systems for controlled delivery of drugs to the body.

In this respect, a polymer or a polymer covalently attached to a drug can have biological activity. Chapter 1 of this introductory section of the book reviews the different types of polymers that effect a biological response in the body. These include polycations, polyanions, and polynucleotides that function as polymeric drugs. Polymers that are conjugated to drugs and have activity are also described as are polymeric prodrugs and targeted polymeric drugs.

Most of the polymeric delivery systems currently being used commercially consist of an established drug physically combined with a biocompatible polymer. Chapter 2 reviews the different classes of polymers used as matrices for drug delivery and the properties of those polymers which make them suited for a particular method of drug delivery. Water-soluble, biodegradable, and nonbiodegradable polymers are discussed and examples of each which are used for drug delivery are described.

Another important method of drug delivery is by means of liposomes. Chapter 3 in this section of the book discusses the different classes of liposomes, their preparation, characterization, and unique properties for drug delivery. Liposomes that release their drug contents at specific temperatures or can be targeted to specific sites are included in this review.

The last paper in this section, Chapter 4, outlines some of the considerations that need to be made in designing any drug delivery system. These include the route of administration and the interaction of the drug with the biological system. By understanding these aspects of drug delivery, it is hoped that new methods and materials can be developed that will lead to more specific and successful controlled delivery of drugs.

Chapter 1

Biologically Active Polymers

Raphael M. Ottenbrite

Department of Chemistry, High Technology Materials Center, Virginia
Commonwealth University, Richmond, VA 23284

Interest continues to grow in polymers that have inherent biological
activity or polymers which are covalently attached to well known drugs.
This paper provides an introduction to these types of materials and a
general review of the different types of polymeric drugs, polymeric drug
conjugates, polymeric prodrugs, and targeted polymeric drugs which can
be classified as biologically active polymers.

Drug delivery systems are ideally devised to disseminate a drug when and
where it is needed and at minimum dose levels. Polymeric drugs and delivery
systems provide that possibility through several different approaches. Polymeric
drugs, polymeric drug conjugates or drug carriers, polymeric prodrug systems,
bioerodible matrices, diffusion through membranes or from monolithic devices, and
osmotic pumps are all drug delivery options. This paper will review some of the
materials and approaches to using polymers as drugs themselves, conjugated to well
known drugs, prodrug systems, and targeted drug carriers.

Polymeric Drugs.

Polymeric drugs are macromolecules that elicit biological activity (1). Many
synthetic polymers are biologically inert. However, some exhibit toxicity, while
others exhibit a wide range of therapeutic activities. There are three kinds of
polymer drugs: polycations, polyanions, and polynucleotides.

Polycationic polymers. These are macromolecules that have electropositive groups
attached to the polymer chain or pendant to the chain. These materials are active
against a number of bacteria and fungi. However, due to their inherent toxicity to
animal species through their destructive interaction with cell membranes they are

0097–6156/91/0469–0003$06.00/0

only used topically (2). Recent reports indicate that they can enhance cellular antigen uptake and exhibit antitumor activity against Ehrlich carcinoma.

Polyanionic polymers. Polyelectrolytes with negative charges on the polymer can also function as drugs. These polymers are much less toxic. Both natural polyanions, such as heparin and heparinoids, and synthetic polyanions such as, poly(acrylic acid), exhibit a variety of biological activity (3). For example, poly(divinyl ether-co-maleic anhydride) exhibits antiviral, antimicrobial and antifungal activity. It is best known for activity against a number of animal tumor models most notable Lewis lung carcinoma, Ehrlich carcinoma, and Friend leukemia. One mechanism of activity is the activation of macrophages and augmentation of natural killer cell activity.

Other polycarboxylates, such as Carbetimer, poly(maleic anhydride-co-cyclohexyl-1,6-dioxepin) and poly(maleic anhydride-co-4-methylpentenane), elicit a similar broad spectrum of activity.

Polyanionic polymers can enter into biological functions by distribution throughout the host and they behave similar to proteins, glycoproteins and polynucleotides which modulate a number of biological responses related to the host defense mechanism. These are enhanced immune responses, and activation of the reticuloendothelial system (RES) macrophages.

The synthetic polymer which has received the most interest is divinyl ether-maleic anhydride copolymer, commonly referred to as pyran copolymer due to the tetrahydropyran ring which forms during polymerizaton. In the literature it is also referred to by the acronym DIVEMA (divinyl ether-maleic anhydride copolymer) and, more recently, as MVE (maleic anhydride-vinyl ether copolymer). Pyran was first reported by Butler in 1960 and submitted to the NIH screen (4). Pyran showed significant activity and was designated as NSC 46015 by the National Cancer Institute. It has been under investigation for use in cancer chemotherapy and has also exhibited a wide range of other biological activities. Pyran has been reported to have interferon inducing capacity and to be active against a number of viruses including Friend leukemia, Rauscher leukemia, Maloney sarcoma, polyoma, vesicular stomatitis, Mengo and encephalomyo-carditis.

A number of investigations have clearly demonstrated that the structure and molecular weight of synthetic polyanions are directly related to biological activity and toxicity. Breslow showed that the acute toxicity caused by pyran in mice increased with increased molecular weight (5). Biological activity data of pyran fractions (2,500-32,000) indicated that molecular weights up to 15,000 stimulated RES whereas higher molecular weight fractions suppressed RES, resulting in biphasic response. It was found that the level of serum glutamic pyruvate transaminase, which is a measure of liver damage, also increased with molecular weight, as did inhibition of drug metabolism and sensitization to endotoxin. However, the activities against Lewis lung and Ehrlich ascites tumor were shown to be independent of molecular weight.

Recently, considerable evidence has emerged that implicates the macrophage as a major effector of tumor cytotoxicity and/or cytostasis. Both synthetic reticuloendothelial stimulants such as poly(acrylic acid-*alt*-maleic anhydride), pyran and poly(riboinosinic-polycytidylic acids)s as well as biologic reticuloendothelial stimulants are known to enhance macrophage function as well as to induce resistance of tumor growth. Moreover, macrophages from animals treated with polyanionic stimulants have been demonstrated to be cytostatic and/or cytotoxic for tumor cells while demonstrating quantitatively less cytotoxicity for normal cells (6).

Presently there are two polyanionic polymers, MVE-2 and Carbetimer, that have undergone considerable clinical evaluation as antitumor agents.

Another important family of anionic polymeric drugs are the heparins and heparinoids which have mainly sulfonate groups. These have been used primarily as anticoagulants but also exhibit antimitotic effects. Recently extrasulfated heparin derived from a heparine fraction was reported to have excellent anti-clotting and anti-Xa activity (7). New fluorescent derivitized heparine molecules, without alteration of their bioactivity, are available for bioevaluations such as the von Willebrand Factor during surgery (8).

More recently chitosan polymers which are derivatives of chitin materials have evoked interest due to their bioactivity and biodegradability. For example, *N*-carboxybutyl chitosan has been show to effectively promote wound healing (9). Acetate, and butyrate derivatives of chitosan have decreased blood clotting time significantly (10).

Polynucleotides. Polynucleotides are potent interferon inducers. A mismatched, double-stranded synthetic polyribonucleotide ampligen and the double-stranded acids, polyadenylic-polyuridylic acid and polyinosinic-polycytidylic acids have been widely studied for cancer therapy(11). Although these materials elicit excellent activity with murine rodents, therapeutic effects are dramatically decreased within primates.

Since nucleic acids and enzymes play such a large role in chromosome replication during mitosis, a considerable amount of research has been conducted in this area to control viruses. On the molecular level, analogues of nucleic acids are capable of forming complexes with adenine, cytosine, uracil, thymine, and guanine. Through complexation, these nucleic acid analogues are potential inhibitors of biosyntheses that require nucleic acids as templates.

The polyvinyl analogues of nucleic acids are one of the few polymers that have been tested in living systems to determine their bioeffects. The most thoroughly investigated polymer is poly(vinyladenine), which has been reported to be effective against viral leukemia, chemically induced leukemia, and infections caused by other viruses (12). The inhibition of viruses by the complexation of nucleic acids with their polymer analogues is apparently very specific. For example, poly(9-vinyladenine) inhibits RNA viral replication through the reverse transcriptase step, whereas poly(9-vinylguanine) is ineffective against viruses. These materials still have clinical interest.

Polymeric Drug Conjugates or Drug Carriers.

Many potentially effective drugs cannot be used because of host elimination or rapid metabolization. In some instances this can be amended by using drug-polymer conjugates (*13,14,15,16*).

The drug is covalently bonded to an appropriate polymer carrier. These large molecules diffuse more slowly and are adsorbed at distinct pharmacological interfaces. Consequently, polymer-drug conjugates can prolong therapy at sustained dosages. The major attributes of polymeric drug carriers are their depot effects, unique pharmacokinetics, body distribution, and pharmacological efficacy.

Most medications are micromolecular in size and are relatively free to diffuse throughout the biological system. Consequently, drugs have been inherently difficult to administer in a localized, concentrated mode within the primary target tissues and organs. Polymers, however, diffuse slowly and are often absorbed at interfaces, the attachment of pharmaceutical moieties can produce a biopolymer with distinct pharmacological behavior. These polymeric drug carriers have desirable properties such as sustained therapy, slow release, prolonged activity, and drug latentation.

A model for pharmacologically active polymer drug carriers, similar to that shown in Figure 1, has been developed (*17*). In this schematic representation, four different groups are attached to a biostable or biodegradable polymer backbone. One group is the pharmacon or drug, the second is a spacing group, the third is a transport system, and the fourth is a group to solubilize the entire biopolymer system. The pharmacon is the entity that elicits the physiological response. It can be attached permanently by a stable bond between the drug and the polymer, or it can be temporarily attached and removed by hydrolysis or by enzymatic processes. The transport system for these soluble polymer drug carriers can be made specific for certain tissue cells with homing or targeting moieties such as pH-sensitive groups or receptor-active components, such as antibody-antigent recognition. Solubilizing groups (such as carboxylates, quaternary amines, and sulfonates) are added to increase the hydrophilicity and solubility of the whole macromolecular system in aqueous media while nonpolar groups enhance the hydrophobic character and solubility in lipid regions.

The specificity of polymer pharmacokinetics is related to the molecular size, which alters the transport rate across compartmental barriers (*18*). By controlling the molecular weight of a polymer drug carrier, it is possible to regulate whether the drug passes through blood-membrane barriers, is excreted by the kidney, or is accumulated in the lymph, spleen, liver, or other organs. The fundamental macromolecular transport theory of biopolymers through tissues has been successfully applied to the design, fabrication, and prediction of in vivo performance of polymer drug-delivery systems. Consequently, based on the molecular weight, conjugate polymer drug carriers have been developed to perform more specifically than the drug alone. Many other variables, such as composition of the polymer chain, structure, polyelectrolytic character, and solubility, can also effect polymer behavior.

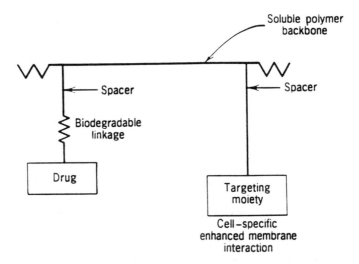

Figure 1. Model for polymer drug carrier.

Probably the most promising polymeric drug carrier system involves polysaccharide molecules. These are natural polymers and are often biodegradable to products that are useful to the host or easily eliminated by the host. Dextrans have been the most extensively used polysaccharide for macromolecular prodrug preparations (19). These materials are biocompatible and the in vivo fate is directly related to their molecular weight. Moreover these macromolecules can be easily targetted to the hepatocytes with D-mannose or L-fucose (20).

Another polysaccharide system that has received considerable interest is the chitosans which are water soluble derivatives of chitin. These materials appear to be very biocompatible and degradable and so are potentially excellent candidates as polymeric drug systems (21).

Recently poly(amino acids) as polymeric drug carriers has increased in use as they form water soluble polymers and hydrophilic gels. Both homopolymers and copolymers have been used. Since they are similar to naturally occuring proteins, the potential biodegradability of poly(amino acids) offers an advantage over the nondegradable synthetic carbon-carbon backbone polymers. The homopolymers of poly(alpha-amino acid)s are usually nonimmunogenic while copolymers with two or more amino acids can be immunogenic. Several poly(amino acid) drug conjugates that have been prepared include poly(L-glutamic acid)-adriamycin and poly (L-lysine)-methotrexate (22).

Polymeric Prodrugs. Polymeric prodrugs are designed to protect against rapid elimination or metabolization by adding a protective polymer to the therapeutic material. Therapeutic activity is usually lost with this attachment and reinstated with the removal of the protective group. The protective group is designed to be easily removed, usually by hydrolysis. Prodrugs are readily prepared with polymers similar to a polymer drug carrier.

Temporary attachment of a pharmacon is necessary if the drug is active only in the free form. A drug which is active only after being cleaved form the polymer chain is called a prodrug. Temporary attachment usually involves a hydrolyzable bond such as an anhydride, ester, acetal, or orthoester. Permanent attachment of the drug moiety is generally used when the drug exhibits activity in the attached form. The pharmacon is usually attached away from the main polymer chain and other pendent groups by means of a spacer moiety that allows for more efficient hydrolysis. For example, catecholamines are ineffective when bonded directly to poly(acrylic acid) but affect heart rates and muscle contractions when attached away from the backbone of the macromolecular carrier. Similarly, isoproterenol elicits a pharmacologic response only when coupled to a polymer by a pendent group.

One of the most successful conjugate polymer systems was developed by Duncan and Kopecek (23). The polymer carrier used in their system is poly [N(2-hydroxypropyl) methacrylamide] a biocompatible polymer that was originally developed as a plasma extender. They have evaluated a number of polymer conjugated drugs for cancer chemotherapy with interesting results. The attachment of the drug is through a peptidyl spacer pendent to the polymer backbone. These peptides links are stable in aqueous media but are readily hydrolyzed intracellularly by lysosomal enzymes. This method provided effective cell specific delivery. Duncan and Kopecek were also able to achieve organ specific delivery to the hepatocytes with galactosamine; to the T-lymphocytes with anti-T cell antibodies; and to leukemia cells with fucosylamine (*24*).

Targeted Polymeric Drugs. Polymer drug targeting to a specific biological site is an enormous advantage in drug delivery because only those sites involved are affected by the drug. This precludes the transport throughout the body, which can elicit serious side effects. Ideally, a targetable drug carrier is captured by the target cell to achieve optimum drug delivery while minimizing the exposure to the host.

Fluid-phase uptake of macromolecules by cells in general is a slow process, and most administered macromolecules are eliminated from the host before any significant cellular uptake takes place. If, however, the macromolecule contains a moiety that is compatible with a receptor on a specific cell surface, then the macromolecule is attracted to the cell surface and the uptake is enhanced. This maximizes the opportunity for specific-cell capture. This type of cell-specific targeting has been developed; to hepatocytes, with galactosamine; to T lymphocytes, with anti-T cell antibodies; and to mouse leukemia cells, with fucosylamine and other biomolecules.

The discovery of monoclonal antibodies and combining them with polymeric prodrugs is the newest approach to overcome the lack of selectivity for disposition in target tissue (*23*). Recently the selectivity of antibody-targeted polymeric anthracycline antibiotics to T lymphocytes was accomplished (*25*). In addition decreased immunogenicity of proteinaceous conjugates with IgG and human transferrin has been reported (*26*).

The ultimate fate of drugs and their metabolites is a major concern. If they are not cleared in a reasonable time, they could promote undesirable side effects. Polymer drug carriers are usually nonbiodegradable, and if their size is greater than 40,000 daltons, they could accumulate in the host with the potential of future unwanted effects.

The applications of targeted drug carrier systems has not been fully realized. The utilization of the total concept, which includes a biocompatible drug-carrier with selective cell targeting, controlled-drug release, and biodegradability, will provide highly potent drug administration systems for the future.

Summary

Polymeric drugs and drug delivery technology has emerged during the 1980's to be a commercially attractive method for administering drugs in a controlled and effective mode. The technique is capable of effecting sustained release of a bioactive substance with optimum response, minimum side-effects, and prolonged efficacy. Recently success has been achieved in targeting drugs to a specific organ or site. The 1990's should prove even more exciting with more clinical activity and new polymer designs.

Literature Cited

1. *Anionic Polymeric Drugs*; Ottenbrite, R. M.; Donaruma, L. G.; Vogl, O., Eds.; Wiler-Interscience: New York, NY, 1980..
2. *Polymeric Drugs*; Donaruma, L. G.; Vogl, O., Eds.; Academic Press, New York, NY, 1978.
3. Ottenbrite, R. M. In *Anticancer and Interferon Agents*; Marcel Dekker: New York, NY, 1984.
4. Butler, G. B. *J. Poly. Sci.* **1960**, *48*, 279.
5. Breslow, D. S. *Pure Appl. Chem.* **1976**, *46*103.
6. Ottenbrite, R. M.; Kuus, K.; Kaplan, A. M. *J. Macromol. Sci.* **1988**, *A25*, 499.
7. Braud, C.; Vert, M.; Petitou M. *J. Bioact. Comp. Polymers* **1989**, *4*, 269.
8. Suda, Y.; Sumi, M.; Sobel, M.; Ottenbrite, R. M. *J. Bioact. Comp. Polymers* **1990**, *5*, 412.
9. Muzzarelli, R. A. A. et al. *J. Bioact. Comp. Polymers* **1990**, *5*, 396.
10. Dutkiewic, J. et al. *J. Bioact. Comp. Polymers* **1990**, *5*, 293.
11. Levy, H. B. *J. Bioact. Comp. Polymers* **1986**, *1*, 348.
12. Takamoto, K; Inaki, Y. *Functional Monomers and Polymers*; Marcel Dekker, Inc.: New York, NY; 1987; 149.
13. Dumitriu, S.; Popa, M.; Dumitriu, M. *J. Bioact. Comp. Polymers* **1988**, *3*, 243.
14. Dumitriu, S.; Popa, M.; Dumitriu, M. *J. Bioact. Comp. Polymers* **1988**, *3*, 403.
15. Dumitriu, S.; Popa, M.; Dumitriu, M. *J. Bioact. Comp. Polymers* **1989**, *3*, 57.
16. Dumitriu, S.; Popa, M.; Dumitriu, M. *J. Bioact. Comp. Polymers* **1989**, *3*, 151.
17. Ringsdorf, H. *J. Polym. Sci.* **1975**, *51*, 135.
18. Johnson, P.; Lloyd-Jones, J. G. *Drug Delivery Systems*; Ellis Horwood Ltd.: 1987.
19. Larsen, C. *Adv. Drug Delivery Rev.* **1989**, *3*, 103.
20. Schwartz, A. L. *CRC Crit. Rev. in Biochem.* **1984**, *16*, 207.
21. Muzzarelli, R. A.; Weckz, A. M.; Filippini, O.; Lough, C. *Carbohydr. Polymers* **1989**, *11*, 307.
22. Rihova, B. *CRC Crit. Rev. in Therapeutic Drug Del. Systems* **1985**, *1*, 311.
23. Kopecek, J. *J. Bioact. Compat. Poly.* **1988**, *3*, 16,
24. Duncan, R.; Seymour, L. W.; Ulbrich, K.; Kopecek, J. *J. Bioact. Compat. Poly.* **1988**, *3*, 4.
25. Rihova, B.; Strohalm, J.; Plocova, D.; Ulbrich, K. *J. Bioact. Compat. Polymers* **1990**, *5*, 249.
26. Flanagan, P.A.; Duncan, R.; Rihova, B.; Subr, V.; Kopecek, J. *J. Bioact. Compat. Polymers* **1990**, *5*, 151.

RECEIVED April 8, 1991

Chapter 2

Polymeric Matrices

Richard L. Dunn

Atrix Laboratories, Inc., 2579 Midpoint Drive, Fort Collins, CO 80525

There are many polymers which have been used as physical matrices for controlled delivery of drugs. In this paper, these polymers are separated into water-soluble, biodegradable, and nonbiodegradable materials. A description of each class of polymers is presented. Examples of polymers from each class that have been used as drug delivery matrices and the criteria for their selection are included in this general review.

One of the simplest methods to obtain sustained delivery of a biologically active agent is to physically mix or blend it with a polymer. Because the drug is now encapsulated or trapped within the polymer, it is not as readily available to the biological system as when it is used alone. The polymer has to dissolve or disintegrate before the drug can be released or else the drug has to dissolve or diffuse from within the polymeric matrix. In either case, the release of the drug to the physiological environment is extended over a much greater time than if the drug is used in its native form.

In these types of controlled delivery systems, the drug is not chemically attached to the polymer, it is simply encased within the polymer matrix. Because there are no chemical interactions between the drug and the polymer, the drug remains in a biologically active form, and can exert its effect upon the body as soon as it is released from the polymeric matrix. The polymer is strictly a vehicle for delivering the drug to the body. Obviously, the major advantage of this type of drug delivery system is that the drug in the polymeric matrix is unaltered; therefore its absorption, distribution, metabolism, and excretion after being released from the polymer is the same as that of the native drug. Moreover, its biological effect when released from the polymeric matrix is the same as that when used alone and can be predicted based upon its release rate.

Because these types of polymeric matrix systems are the simplest to design and the easiest to obtain approval by the Food and Drug Administration, they have been the most extensively studied in the past two decades. Numerous polymers have been evaluated for these types of drug delivery systems and although it would be impractical to present each of these polymers and its specific application to drug delivery, this chapter will review in general the types of polymers used as matrices for drug delivery (1-4).

The polymers used as matrices for drug delivery can be classified as three basic types. These are water-soluble polymers, biodegradable polymers, and nonbiodegradable polymers. Water-soluble polymers are typically used for short-term (several hours to several days) drug delivery because of their quick dissolution in biological fluids. Their solubility in body fluids is a result of functional groups such as hydroxyl, amine, and carboxylic acid that are present on the polymer chain. These polymers do not have to undergo any chemical degradation to become water soluble; they simply have to become hydrated, ionized, or protonated to lose their form and dissolve within the body. Examples are polyethylene glycol, polyethyleneimine, and polyacrylic acid.

Biodegradable polymers are those that require some chemical reaction or alteration in the body before they become water-soluble. Because the degradation involves some type of chemical reaction such as hydrolysis which usually occurs over an extended period of time, these polymers can be used for longer-term drug delivery. Extensive research has produced biodegradable polymers with degradation times ranging from several days to more than one year; thus these polymers afford a wide range of applications for drug delivery.

The third general class of polymers used as matrices for drug delivery are the nonbiodegradable polymers. Because these polymers are essentially inert in the body, they can be used for applications in which extended drug delivery is required. Also, because of their inertness, they will remain in the body indefinitely after the drug is depleted unless they are removed.

The decision to use a polymer from one of these three general classes of polymeric matrices depends upon the route of administration, the type of drug used, the amount of drug to be delivered, and the duration of release required. First to be considered is the route of administration. Just about any body opening or cavity can be considered a potential site for drug administration. Thus oral, nasal, buccal, ocular, and rectal are common routes of drug administration. In addition, drugs can be delivered parenterally to the body by means of injection such as intraveneously, intramuscularly, intraperitoneally, or subcutaneously. Although these methods are often not as convenient and painless as the body opening methods, they do provide a quick and effective route of administration for many drugs. Drugs can also be applied to the surface of the body. Topical applications of drugs have been used extensively for centuries to treat topical infections and skin diseases. More recently, the topical application of drugs combined with vehicles to allow passage of the drug through the skin has given a new method to deliver systemic drugs which has been termed transdermal. Each route of administration has its own advantage and disadvantages and the selection of the most appropriate route depends upon the type of drug to be given, the quantity needed for clinical efficacy, and the duration of administration.

For instance, if the drug is unstable in the presence of gastric or intestinal fluids as are many proteins or if it is quickly metabolised during passage through the liver, then it will have to be delivered parenterally. However, if large quantities of a drug are required each day to effect biological activity, then intramuscular, intraperitoneal, and subcutaneous injections would be limited as a result of the injection volume. Moreover, if a drug needs to be given over an extended time frame, the problems with injection encourage the selection of an oral route if possible. Obviously, sometimes the preferred route is not feasible as is the case with the continued daily injections of insulin by diabetics. Sometimes, even though a particular route of administration is appropriate for the drug to be delivered, the perceived inconvenience to the patient often precludes widespread use such as with rectal and buccal administration of drugs in the United States.

The route of administration is a major consideration in the type of polymer selected for the drug matrix. Although the gastrointestinal tract can accommodate nonbiodegradable/water-insoluble polymers and pass them through, water-soluble polymers are preferred in oral drug delivery so that no retrieval of the polymer is ever contemplated. The same condition holds for most of the other mucosal tissue administration such as ocular, nasal, and rectal. Most of the drugs delivered to these areas utilize water-soluble polymers with fairly short durations of release. However, there are nonbiodegradable polymers used for ocular drug administration. In these cases, a device fabricated from the polymer and drug is inserted into the eye and removed when the drug is depleted. Similar types of devices fabricated from nonbiodegradable polymers are also used in subcutaneous injections where they can be easily removed at the end of treatment; however, biodegradable polymers are preferred for the intramuscular and intraperitoneal injections where retrieval of a nonbiodegradable polymer is less feasible.

The nature of the drug being delivered also influences the selection of the polymer used for the drug delivery matrix. If the drug is highly acidic or basic, a polymer has to be selected that is compatible with the drug and one that does not lead to reactions which will alter either the drug or the polymer. This is especially true with biodegradable polymers in which the drug may catalyze the degradation. Also if a drug is very unstable under aqueous conditions, a more hydrophobic and less water-soluble polymer is selected depending upon the duration of release.

The amount of drug to be delivered and the duration of release also influence the polymer selection. The amount of the drug to be delivered directly affects the route of administration which in turn affects the type of polymer needed. The duration of release has a direct effect upon the polymer selection. As discussed previously, water-soluble polymers are used mainly for short-term drug delivery, biodegradable for intermediate lengths of time, and nonbiodegradable for the more extended time periods. However, all three types can be used for short-term delivery if the system is properly designed so that the rate of drug delivery is independent of the polymer dissolution or degradation. The following sections will attempt to describe the most commonly used polymers in each class of polymeric matrices and some of their applications to drug delivery.

Water-Soluble Polymers.

The water-soluble polymers with the most extensive use as drug delivery matrices are those that contain hydroxyl, ethyleneoxy, amine, and carboxy acid groups. These polymers dissolve in the body as a result of hydration such as those with hydroxyl or etheneoxy groups or because of ionization or protonation such as those with amine or carboxylic acid groups. Table I lists some of the more widely used water-soluble polymers. Many of these are cellulose-based polymers and all are commonly used in the preparation of tablets and capsules of drugs given orally. Although these polymers are water-soluble, their rates of dissolution vary over a wide range. Even with a certain polymer, the rate of water dissolution can be changed rather significantly by changing the molecular weight of the polymer. A prime example is polyethylene glycol in which a polymer with a molecular weight of 1000 Daltons dissolves much faster than one with a molecular weight of 4 million Daltons. Careful selection of the polymer with the desired dissolution rate combined with control of the tabletting or capsule-forming process can lead to a number of formulations with different drug release profiles and rates.

With tablets, several mechanisms of drug release may be occurring at the same time. First, there is a dissolution of the drug from within the polymeric matrix. The size and shape of the tablet will affect the surface area exposed to water. This parameter, the drug content of the tablet, and the solubility of the drug in body fluids will determine its rate of release by dissolution. Moreover, the tablets may have additives which aid in disintegration. Disintegration of the tablet exposes more surface area and speeds the drug dissolution. Another factor can be the diffusion of the drug itself through the hydrated polymer to the surface and then dissolution in the fluid. Consequently, a number of different parameters can be evaluated with tablets to give the desired rate of release (5,6).

With capsules, the polymer usually completely encloses the drug in the form of a membrane. The rate of dissolution of the polymer and the thickness of the membrane then determine the time at which the drug is exposed fully to body fluids. Of course, the drug can also diffuse through the polymer membrane to the surface with subsequent dissolution. In this instance, the rate of release is more constant.

Polymers that dissolve in water as a result of protonation or ionization can also be used to control drug delivery especially in the gastrointestinal tract. Many drugs are either destroyed by the acidic environment or else poorly absorbed in the stomach. Polymers such as the half esters of maleic anhydride can be applied to tablets or capsules containing these drugs. At low pH's, these polymers are not ionized and the ester group provides enough hydrophobicity to prevent dissolution. Therefore, these polymers do not dissolve in the stomach. But at the higher pH's in the small intestine, enough acid groups are ionized such that the polymer becomes water soluble. By using mixtures of these polymers as enteric coatings, the disintegration and dissolution time in the intestine can be prolonged. Similarly, drugs that need to be absorbed in the stomach can be coated with amine-containing polymers. The protonation of these polymers under acidic conditions makes them water soluble (7,8).

Water-soluble polymers can also be used as aqueous solutions for drug delivery. Although the polymer is already dissolved, its increase in viscosity of the drug solution causes the drug to be retained somewhat longer in the desired application. This technique is common with ocular, nasal, and oral applications of drug solutions.

A number of the water-soluble polymers also have adhesive properties which are being extensively evaluated for drug delivery (9). These polymers will adhere to the mucous coating in the gastrointestinal tract, the nose, and the mouth to delay passage and sustain drug release. Those polymers with the best adhesive properties are those with hydroxyl and carboxyl groups. Table II lists some of the bioadhesive polymers and their adhesive properties.

Biodegradable Polymers.

The term "biodegradable polymers" as used in this paper denotes water-insoluble polymers which by means of a chemical reaction in the body are converted slowly to water-soluble materials. Several methods for achieving this change in the body are known. First, the polymer can have a side chain substituent which undergoes hydrolysis in the body to produce hydroxyl, carboxyl, or other hydrating groups. These groups then act to make the entire polymer water-soluble. Another approach is to crosslink a water-soluble polymer with a hydrolyzable crosslinking agent. Once crosslinked, the polymer is insoluble. But when placed in the body, the crosslinking group is hydrolyzed or degraded to give a water soluble polymer. A third and more frequently used technique is to use water-insoluble polymers that contain hydrolyzable functional groups directly in the polymer chain. As these groups in the chain are hydrolyzed, the polymer chain is slowly reduced to shorter and shorter chain segments which eventually become water-soluble. The main advantage of this latter group of polymers is that extremely high-molecular-weight polymer with good mechanical properties can be used and subsequently eliminated from the body when the polymer chain has been reduced to water-soluble fragments. With the other type of biodegradable polymers, the polymer chain of the water-soluble segment must be less than approximately 50,000 Daltons to be eliminated from the body as the carbon-carbon backbone of the polymer chains will not undergo degradation to shorter chain lengths. Table III lists examples of these three categories of biodegradable polymers.

Although there are a number of examples where the first two categories of biodegradable polymers have been evaluated for drug delivery, generally they have not received as much attention as the third. Some of the reasons may be the unavailability of commercial polymers in this area, the problems of controlling the hydrolysis rate of the side chain groups as well as preventing reaction of these groups with the drug itself, and the difficulties in providing enough crosslinking to make the polymer insoluble but also degradable within a reasonable length of time.

Table I. Examples of Water-Soluble Polymers Used as Drug Delivery Matrices

Polyethylene glycol	Dextran
Poly(vinyl alcohol)	Sodium alginate
Poly(vinyl pyrrolidone)	Poly(acrylic acid)
Poly(2-hydroxyethyl methacrylate)	Poly(methacrylic acid)
Poly(acrylamide)	Poly(maleic acid half esters)
Hydroxypropyl cellulose	Poly(sodium styrene sulfonate)
Hydroxypropylmethyl cellulose	Poly(dimethylaminoethyl methacrylate)
Sodium carboxymethyl cellulose	Poly(vinyl pryidine)
Gelatin	Cellulose acetate $N,N,$-diethylaminoacetate
Starch	

Table II. Examples of Polymers with Bioadhesive Properties

Polymer	Mean % adhesive Force
Sodium carboxymethyl cellulose	192.4
Poly(acrylic acid)	185.0
Poly(methyl vinyl ether-co-maleic anhydride	147.7
Poly(ethylene oxide)	128.6
Methylcellulose	128.0
Sodium alginate	126.2
Starch	117.2
Gelatin	115.8
Poly(vinyl pyrrolidone)	97.6
Poly(ethylene glycol)	96.0
Hydroxypropylcellulose	87.1

Adapted from Ref. 9

Table III. Examples of Biodegradable Polymers Used in Drug Delivery

Category I

Poly(maleic anhydride copolymers)

Category II

Gelatin-formaldehyde
Acrylamide - N, N'-methylenebisacrylamide
N-vinyl pyrrolidone - N, N'-methylenebisacrylamide
Fumaric acid/polyethylene glycol - N-vinyl pyrrolidone
Fumaric acid/diglycolic acid - N-vinyl pyrrolidone
Fumaric acid/ketomalonic acid - N-vinyl pyrrolidone
Fumaric acid/ketoglutaric acid - N-vinyl pyrrolidone

Category III

Polylactic acid	Polyanhydrides
Polyglycolic acid	Polyorthoesters
Polycaprolactone	Poly(amino acids)
Polyhydroxybutyrate	Psuedopolyamino acids
Polyhydroxyvalerate	Polyphosphazenes

In addition, the incorporation of a drug within the crosslinked polymer can present some problems (*10*). For these reasons, more attention has been devoted to the third category of biodegradable polymers. However, with the advent of high molecular weight drugs such as those produced by genetic engineering these two categories of biodegradable polymers, especially the crosslinked polymers, may receive new attention in the future.

The third category of biodegradable polymers in which the polymer chain is hydrolyzed into shorter and shorter segments has been extensively studied for drug delivery applications. One of the most versatile and widely used synthetic biodegradable polymers are the aliphatic polyesters prepared from lactic and glycolic acids. Both homopolymers and copolymers of these two monomers have been synthesized, characterized, and evaluated as drug delivery matrices. Because these polymers were first utilized as sutures, their biocompatibility is well known and their approval by the Food and Drug Administration for use as drug delivery matrices presents no major problems. Moreover, the more recent commercial availability of these polymers and the ability to accurately control their biodegradation rates has made these biodegradable polymers the first choice for pharmaceutical companies developing biodegradable controlled release systems for their drugs. Several excellent reviews have been published describing these polymers, their synthesis, characterization, and fabrication into microcapsules, films, fibers, and rods for controlled release of drugs ranging from low molecular weight antibiotics and steroids to high molecular weight proteins such as vaccines and growth hormones (*7,8,10,11,12*). Although the majority of drug delivery systems developed with these polymers have been injected or implanted either subcutaneously or intramuscularly, there are studies showing their use in nasal and oral administration (*13*). These polymers represent a safe and versatile matrix for drug delivery and there is no question that a series of controlled release drug products based upon these products will be commercialized in the next several years.

Aliphatic polyesters based on monomers other than α-hydroxyalkanoic acids have also been developed and evaluated as drug delivery matrices. These include the polyhydroxybutyrate and polyhydroxyvalerate homo- and copolymers developed by Imperial Chemical Industries (ICI) from a fermentation process and the polycaprolactones extensively studied by Pitt and Schindler (*14,15*). The homopolymers in these series of aliphatic polyesters are hydrophobic and crystalline in structure. Because of these properties, these polyesters normally have long degradation times in vivo of 1-2 years. However, the use of copolymers and in the case of polycaprolactone even polymer blends have led to materials with useful degradation times as a result of changes in the crystallinity and hydrophobicity of these polymers. An even larger family of polymers based upon hydroxyaliphatic acids has recently been prepared by bacteria fermentation processes, and it is anticipated that some of these materials may be evaluated for drug delivery as soon as they become commercially available.

Two other series of biodegradable polymers that depend upon chain degradation are the polyanhydrides and the polyorthoesters. Both of these polymers contain hydrophobic units linked together along the polymer chain by functional groups very susceptible to hydrolysis. Because of their hydrophobic character, only the outer layers are penetrated with water when implanted in the body. Exposure of the outer layer of polymer to water causes fast hydrolysis of the water-labile linkages and subsequent degradation and loss of the outer layer of polymer. As the outer layer is lost, another layer of polymer is exposed to water and the process continues such that the polymers are slowly eroded from the surface. Such polymers are termed bioerodible and the rate of hydrolytic degradation at the surface of the polymer is much faster than the rate of water permeation into the polymer matrix.

The polyanhydrides developed by Langer are prepared from the diacid monomer, bis (*p*-carboxyphenoxy) methane (*16*). This hydrophobic monomer leads to polymers with low water uptake or permeation. However, the anhydride linkage is quite susceptible to hydrolysis and experiments have demonstrated that surface erosion of these polymers does occur. However, it has been predicted that complete degradation of the hydrophobic homopolymer will require over three years. Therefore copolymers with sebacic acid and other diacids have been prepared to give shorter degradation times of 2-10 weeks. The degradation rate of the copolymers as well as the homopolymers are dependent upon pH. The higher the pH of the solution to which they are exposed, the faster their degradation rates. In many examples of drug release from these polymers, the rate of drug release has correlated well with polymer degradation or erosion. However in some compression-molded samples, drug release has been faster than polymer erosion possibly as a result of inhomogeneous drug dispersion or a microporous structure. Also, some of the delivery devices fabricated from the copolymers tend to disintegrate and fragment after a certain amount of degradation (*17*). Regardless of these minor problems, the polyanhydrides continue to offer significant promise in the area of drug delivery because of their erosion characteristics, zero-order drug delivery, and apparent lack of toxicity in animals and humans. The ability to protect a drug susceptible to degradation in body fluids and deliver it at a constant rate and in an active form in the body makes this polymeric system potentially useful for a number of proteins and other drugs (*18*).

The polyorthoesters developed by Alza Corporation and Heller at Stanford Research Institute are also bioerodible polymers. However, these polymers contain an orthoester linkage which is more susceptible to hydrolysis under low pH conditions than basic conditions. Control of the degradation rate of these polymers has been achieved by incorporating acidic or basic excipients into the polymer matrix (*19,20,21,22,23*). If long term delivery of a drug is needed, then the polyorthoester may be fabricated into a delivery device containing a basic salt such as magnesium hydroxide. This base stabilizes the entire polymer but as the salt is dissolved away from the surface or neutralized, the stabilization is lost and the outer layers start to degrade. Similarly, if fast degradation of the hydrophobic polyorthoesters is needed, then an acidic excipient such as 9,10-dihydroxy stearic acid or an excipient such as phthalic anhydride that can hydrolyze to give an acidic excipient can be used. Although Heller and other investigators have shown fairly good correlation of drug release with polymer erosion, the correlation tends to become less evident at longer time points. However, a number of drug delivery systems with zero-order release rates have been developed using polyorthoesters, and they appear to offer considerable opportunities for future systems as the polymers become more available and their safety for use in humans is approved by FDA.

Other biodegradable polymers have also been developed and evaluated for drug delivery. These include the poly(amino acids) developed by Sidman and coworkers (3), the pseudopoly(amino acids) investigated by Kohn (24), and polyphosphazenes synthesized by Allcock and coworkers (25). The poly(amino acids) evaluated by Sidman were shown to be potentially useful for delivery of naltrexone in a drug delivery system, but these biodegradable polymers have not been extensively evaluated or commercialized for use in drug delivery. The pseudopoly(amino acids) and the polyphosphazenes are relatively new biodegradable polymers and their use as drug delivery matrices are just now becoming more widely studied. They appear to offer a number of possibilities for drug delivery unavailable with the current polymers.

Nonbiodegradable Polymers.

Nonbiodegradable polymers are those that are essentially inert and do not undergo any chemical change in vivo. Both hydrophobic and hydrophilic polymers may fit this category but neither should be water soluble. Table IV gives a list of the most commonly used hydrophobic nonbiodegradable polymers used for drug delivery. These include the silicones which were one of the first polymers studied by Folkman and Roseman. Silicones are excellent materials for drug delivery because of their biocompatibility, their ease of fabrication into various devices and their high permeability to many drugs. As a result, a number of drug delivery products employing silicones have been commercialized including implants for contraceptive steroids and transdermal devices for nitroglycerin. Both matrix and reservoir type of drug delivery systems have been fabricated with silicones and the release rates accurately predicted and controlled according to Fickian diffusion theory (3).

Ethylene vinyl acetate has also found major applications in drug delivery. These copolymers used in drug release normally contain 30-50 wt% of vinyl acetate. They have been commercialized by the Alza Corporation for the delivery of pilocarpine over a one-week period (Ocusert) and the delivery of progesterone for over one year in the form of an intrauterine device (Progestasert). Ethylene vinyl acetate has also been evaluated for the release of macromolecules such as proteins. The release of proteins form these polymers is by a porous diffusion and the pore structure can be used to control the rate of release (3). Similar nonbiodegradable polymers such as the polyurethanes, polyethylenes, polytetrafluoroethylene and poly(methyl methacrylate) have also been used to deliver a variety of different pharmaceutical agents usually as implants or removal devices.

Equally important as nonbiodegradable polymers for drug delivery are the water-soluble polymers crosslinked by nondegradable (non-hydrolyzable) linkages. Although these polymers are nonbiodegradable and nonsoluble in the body, they do absorb large quantities of water and swell. Typical crosslinking reactions include the use of difunctional vinyl monomers, diisocyanates, glutaldehydye, and γ-

Table IV. Examples of Nonbiodegradable Polymers Used in Drug Delivery

Silicones	Polyisobutylene
Poly(ethylene vinyl acetate)	Cellulose acetate
Poly(methyl methacrylate)	Poly(ethyl methacrylate)
Polyethylene	Poly(butyl methacrylate)
Polyurethanes	

irradiation. If the swelling of the hydrogel is properly controlled, then the polymeric drug delivery device swells at a rate to compensate for the drug lost from the surface. The swelling of the hydrogel increases the surface area of the drug delivery device and more drug should be released. This increase in drug release will counter the effect of decreasing drug release caused by diffusion of drug from the outer portion of the device. If the two rates can be properly controlled, then nearly zero-order drug release may be obtained (26,27).

Hydrogels are used in oral delivery of drugs and as devices in rectal, vaginal, and buccal delivery of drugs. They afford a high degree of compatibility and stability to many drugs, and they may prove to be of great benefit in the future delivery of some proteins and peptide drugs.

Conclusion

Although this limited introduction to polymers used in drug delivery shows that there is an extensive list of biocompatible polymers that can be used, the selection of the most appropriate polymer for any drug delivery systems is dependent upon a number of factors. All of these factors have to be considered when choosing a polymer for delivering a specific drug by a specific route of administration. Delivering either another drug or even the same drug but by a different route of administration may require an entirely different polymer. Too many researchers and companies make the mistake of assuming that if a polymer works with one drug, then it can be extrapolated to work for all drugs. This assumption is incorrect and leads to numerous problems and disappointments in developing drug delivery systems. Because of the differences in drugs and their routes of administration, a wide variety of polymers are needed to fit the many applications. Although many of the commercially available polymers meet the criteria for successful delivery of many drugs, the advent of new drugs that are potent but highly unstable in the body require the development of new polymers specifically designed for this type of application. Because of this demand, new polymers with more controllable properties will continue to be developed and extend the application of polymeric materials to drug delivery.

Literature Cited

1. Wood, D. A. *Crit. Rev. Appl. Chem.* **1984**, *6*, 71.
2. Langer, R. S.; Peppas, N. A. *Biomaterials* **1981**, *2*, 201.
3. Langer, R.; Peppas, N. *Rev. Macromol. Chem. Phys.* **1983**, 61.
4. Ottenbrite, R. M. *Encyclopedia of Polymer Science and Engineering;* John Wiley & Sons, Inc.: 1989, Supplement Vol. 2nd Edition, pp. 164-187.
5. Ganderton, D. In *Drug Delivery Systems;* Johnsten, P., Lloyd-Jones, J. G., Ed.; Ellis Harwood Publishers, Chichester, UK, 1987.
6. Chien, Y. W. *Drug Development and Industrial Pharmacy;* Marcel Dekker, Inc., New York, NY, 1983; Vol. 9, pp. 1291-1330.
7. Heller, J. In *Critical Reviews in Therapeutic Drug Carrier Systems;* Bruck, S. D., Ed. CRC Press, Inc., Boca Raton, FL, 1984, Vol. 1; pp. 39-90.
8. Heller, J. In *Medical Applications of Controlled Release;* Langer, R. S., Wise, D. L., Ed.; CRC Press, Inc., Boca Raton, FL, 1984, pp. 69-101.
9. *Bioadhesion-Possibilities and Future Trends;* Gurney, R.; Junginger, H. E., Eds.; Wissenscheftliche Verlagsgesellachaft mbh, Stultgart Germany, 1990, pp. 13-213.
10. Heller, J.; Helwing, R. F.; Baker, R. W.; Tuttle, M. E. *Biomaterials.* **1983**, *4*, pp. 262.
11. Holland, S. J.; Tighe, B. J.; Gould, P. L. *J. Control. Rel.* **1986**, *4*, 155.
12. Lewis, D. H. In *Biodegradable Polymers as Drug Delivery Systems;* Chasin, M.; Langer, R., Eds.; Drugs and the Pharmaceutical Sciences; Marcel Dekker, Inc., New York, NY, 1990, Vol. 45, pp. 1-41.
13. Eldridge, J. H.; Hammond, C. J.; Meulbrook, J. A.; Stass, J. K.; Gilley, R. M.; Tice, T. R. *J. Control. Rel.* **1990**, 11, 205.
14. Schinder, A.; Jeffcoat, R.; Kimmel, G. L.; Pitt, C. G.; Wall, M. E.; Zweidinger, R. In *Contemporary Topics in Polymer Science;* Pearce, E. M.; Schaefgen, J. R., Eds.; Plenum Publishing Corporation: 1977, Vol. 2; pp. 251-189.
15. Pitt, C. G. In *Biodegradable Polymers as Drug Delivery Systems;* Chasin, M.; Langer, R., Eds.; Drugs and the Pharmaceutical Sciencesw; Marcel Dekker, Inc., New York, NY, 1990, Vol. 45, pp. 71-115.
16. Chasin, M.; Domb, A.; Ron, E.; Mathiowitz, E.; Langer, R.; Leong, K.; Laurencin, C.; Brem, H.; Grosoman, S. In *Biodegradable Polymers as Drug Delivery Systems;* Chasin, M.; Langer, R., Eds.; Drugs and the Pharmaceutical Sciences, Marcel Dekker, Inc., New York, NY, 1990, Vol. 45; pp. 43-70.
17. Leong, K. W.; Brott, B. C.; Langer, R. *J. Biomed. Mat. Res,* **1985**, *5*, 13.
18. Mathiowitz, E.; Langer, R. *J. Control. Rel.,* **1987**, *5*, 13.
19. Heller, J.; Pehnale, W. H.; Helwing, R. F.; Fritzinger, B. K. *Polym. Eng. and Sci.,* **1981**, *21*, 727.
20. Sparer, R. V.; Shik, C.; Ringlisen, D. C. D.; Himmelstein, K. J. *J. Control. Rel.,* **1984**, *1*, 23.
21. Nguyen, T. H.; Higuchi, T.; Himmelstein, K. J. *J. Control. Rel.,* **1987**, *5*, 1.

22. Heller, J.; Perhale, D. W.; Fritzinger, B.K.; Ng, S. Y. *J. Control. Rel.* **1987**, *5*, 173.

23. Heller, J.; Sparer, R. V.; Zentner, G. M. In *Biodegradable Polymers as Drug Delivery Systems;* Chasin, M.; Langer, R., Eds.; Drugs and the Pharmaceutical Sciences; Marcel Dekker, Inc., New York, NY, 1990, Vol. 45; pp. 121-161.

24. Kohn, J. In *Biodegradable Polymers as Drug Delivery Systems;* Chasin M.; Langer, R., Eds.; Drugs and the Pharmaceutical Sciences: Marcel Dekker, Inc., New York, NY, 1990, Vol. 45, pp. 195-229.

25. Allcock, H. R. In *Biodegradable Polymers as Drug Delivery Systems;* Chasin, M.; Langer R., Eds.; Drugs and the Pharmaceutical Sciences: Macel Dekker, Inc., New York, NY, 1990, Vol. 45, pp. 163-193.

26. Korsmeyer, R. W.; Peppas, N. A. *J. Control. Rel.,* **1984**, *1*, 89.

27. Graham, N. B. In *Proceedings of the NATO Advanced Study Institute on Biopolymers;* Piskin E., Hoffman, A. S., Eds.; NATO ASI Series, Series E, No. 106; Martinus Nijhoff Publishers: Dordrecht, The Netherlands, 1986, pp. 170-194.

RECEIVED March 19, 1991

Chapter 3

Liposomes

Naoto Oku[1]

Faculty of Pharmaceutical Sciences, Setsunan University, Hirakata, Osaka, Japan

Liposomes represent an important class of carrier vehicles other than polymers for drug delivery. This paper provides an introduction and general review of liposomes with emphasis on their classifications, their constituent materials, their preparation and characterizaton, and their stability and biodistribution in the body. Liposomes with specific characteristics are also described in this general introduction.

Phospholipids derived from biomembranes when dispersed in water form lipid bilayer vesicles named liposomes which enclose an aqueous space inside the vesicle (Figure 1). Drugs and other substances can be entrapped either in the internal aqueous phase of the liposomes or in the lipid bilayers depending on their hydrophilicity. Liposomes are being used as carrier vehicles for drugs, such as antifungal and anticancer drugs and represent a new drug delivery system. When liposomes are used as drug carriers, it is possible to target the delivery site, enhance and sustain the clinical effects, reduce the toxicity of drugs, and protect drugs from metabolism and immune response (1-7).

Liposomes are formed due to the amphiphilic character of lipids which assemble into bilayers by the force of hydrophobic interaction. Similar assemblies of lipids form microspheres when neutral lipids, such as triglycerides, are dispersed with phospholipids. Liposomes are conventionally classified into three groups by their morphology, i.e., multilamellar vesicle (MLV), small unilamellar vesicle (SUV), and large unilamellar vesicle (LUV). This classification of liposomes is useful when liposomes are used as models for biomembranes. However, when liposomes are used as capsules for drugs, size and homogeneity of the liposomes are more important than the number of lamellars in a liposome. Therefore, "sized" liposomes are preferred. These are prepared by extrusion through a polycarbonate

[1]Current address: School of Pharmaceutical Sciences, University of Shizuoka, Yada, Shizuoka 422, Japan

filter with distinct pore sizes (*8*). Shown in Table I are several kinds of liposomes, which differ in size, lamellar number and preparation method.

Multilamellar vesicles (MLVs) are easily prepared by simple hydration of phospholipids, and are rather stable. MLVs consist of a series of concentric bilayers separated by an aqueous compartment. If MLVs are administered by intravenous injection (iv), they tend to be taken up by reticuloendothelial systems (RES), such as the liver and spleen. MLVs are suited for sustained release of drugs by subcutaneous or intramuscular administration. Small unilamellar vesicles (SUVs) can be prepared homogeneously with diameters of 20-100 nm. SUVs tend to be retained in blood for longer periods when administered iv. Although extremely large molecules cannot be entrapped in SUVs, materials less than 20,000 in molecular weight can be entrapped and at the same trapping efficiency (*15*). SUVs specifically composed of phospholipids with saturated fatty acyl chains (saturated phospholipids), have a tendency to fuse together (*16*). Large unilamellar vesicles (LUVs) have a larger internal volume and can be used to encapsulate huge materials such as bacteriophages (*17*). The diameters of LUVs are 100-1000 nm.

Liposomes and Their Constituent Lipids

Shapes of Lipids. Not all lipids derived from biomembranes prefer to form lipid bilayers. Membrane lipids can be classified into three groups by their shape, namely relative volume of polar head group and the size of the hydrophobic acyl chain (*18*). The first group consists of cylindrical shaped lipids, such as phosphatidylcholine (PC), phosphatidylserine (PS), phosphatidylglycerol (PG), and sphingomyelin (SM). These lipids tend to form stable bilayers in aqueous solution, and therefore, they are designated bilayer forming lipids. Phosphatidylcholines (PC) are most commonly used for making liposomes. PC liposomes are quite stable against the changes in pH or salt concentration.

The second group consists of cone-shaped lipids, such as phosphatidylethanolamine (PE) and cardiolipin (CL) which have a large hydrophobic tail against the polar head groups. These lipids are called nonbilayer lipids, and form a hexagonal II structure as well as bilayer structure depending on the conditions (*19*). Nonbilayer lipids are thought to be important for membrane fusion (*20*). The third group of lipids consists of wedge-shaped lipids. Lyso-phospholipids are classified into this group. Liposomal lipids should be selected according to the purpose. For the purpose of sustained release of encapsulated drugs, bilayer forming lipids are used for making liposomes. On the other hand, nonbilayer lipids are preferable for making fusogenic liposomes.

Charges of Lipids. Natural phospholipids have neutral or negative charges at neutral pH. Charged lipids incorporated into liposomal membrane prevent liposomal aggregation, and expand the aqueous phase between lamellars of MLV resulting in an increase of the trapped volume (*21*). Membrane charge affects the permeability and trapping efficiency of encapsulated drugs or materials. It also affects the biodistribution of liposomes. Stearylamines (positive), PS, PG, and

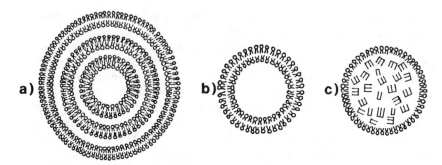

Figure 1. Structure of liposomes and lipid microsphere a). multilamellar vesicle b). unilamellar vesicle c). lipid microsphere. Symbols inside the microsphere indicate di- and tri-acyl glycerol.

Table I. Classification of Liposomes

Abbreviation and Description of Liposomes	References
MLVs Multilamellar vesicles: easy to make, stable	
SUVs Small unilamellar vesicles: 20-100 nm, homogeneous	
LUVs Large unilamellar vesicles: 100-1000 nm, large trapped volume	
GUVs Giant unilamellar vesicles: >1 μm, huge trapped volume, unstable	9, 10
REVs Reverse-phase evaporation vesicles: LUVs prepared by reverse-phase evaporation method, high encapsulation efficiency	11, 12
FTVs Freeze-thaw vesicles: LUVs prepared by freeze-thaw method, high encapsulation efficiency	
VETs Vesicles by extrusion techniques: homogeneous sized vesicles	8
SPLVs Stable plurilamellar vesicles: MLVs without osmotic unbalance	13
MPVs Monophasic vesicles: SPLVs prepared with water-miscible solvent	14

phosphatydic acid (PA) (negative) are usual components of liposomal membranes in 10-20 mole % to endow charges.

Phase Transition of Lipids. Lipids with saturated acyl chains form a solid bilayer (gel phase) at lower temperatures where the mobility of hydrocarbon chains is suppressed. This bilayer undergoes a phase transition with an increase in the temperature, and enters a fluid phase (liquid crystalline phase above the phase transition temperature (T_c). The T_c increases with an increase in fatty acyl chain length, and decreases with unsaturation of the acyl chain. If phospholipids have the same fatty acyl chain, PA, PS, and PE have a higher T_c than PC or PG due to the hydrogen bonding between the polar head groups (*22*). A liposomal membrane is fluid and permeable at a liquid crystalline phase compared with that of a gel phase. Liposomes composed of saturated phospholipids lose barrier function during phase transition (*23*). This character, allows drugs encapsulated in liposomes composed of dipalmitoyl PC (DPPC) and distearoyl PC (DSPC) to be released at warmer target sites (*24*)

Effect of Cholesterol. Cholesterol inclusion into the lipid bilayers composed of DPPC or DSPC, eliminates apparent T_c and reduces permeability at and above the usual T_c. On the other hand, cholesterol inclusion increases packing of fluid bilayer composed of lipids with unsaturated fatty acyl chains. Since cholesterol rich liposomes are stable in plasma, cholesterol is commonly used as a liposomal component.

Preparation of Liposomes

Typical MLVs are prepared by simple hydration of lipids in an aqueous solution with mechanical agitation using a vortex mixer. If the lipids are hydrated without agitation, giant oligo- or uni-lamellar vesicles are produced (*9,25*). Once liposomes are prepared by hydration and then mechanically broken, they reform closed vesicles again with changes in size and lamellar number. SUVs are produced by mechanical breakdown of MLVs using sonication or a French press. Homogeneous sized liposomes are produced by extrusion through a polycarbonate membrane with pores of defined sizes. For example, SUVs are produced by continuous extrusion of MLVs through a 100 nm pore (*8,26*). To make a large quantity of smaller sized liposomes, a microemulsification method has been developed (*27*). Rearrangement of lipid bilayers also occurs by a dehydration-rehydration process, such as freeze-thawing. When a SUV solution is frozen, SUVs fuse together due to ice growth, and rehydrated lipids form LUVs or giant unilamellar vesicles (GUVs) after thawing (*28,29*). Procedures for producing liposomes by simple hydration with or without rearrangement of lipid bilayers are suitable for mass production of liposomes and for liposomal manufacturing.

Other methods for preparing liposomes have been developed, such as hydration of lipids from an organic phase, and detergent removal methods. A lipid bilayer is formed in an aqueous solution of lipids dissolved in an organic solvent when the solvent is removed or diluted. For example, LUVs are formed from a water-in-oil emulsion, followed by removal of solvent (11,12). This technique has a high "trapping" efficiency, and is called the reverse-phase evaporation method. Unilamellar vesicles are formed upon the removal of detergent from mixed micelles composed of detergent and phospholipids by dialysis, gel filtration, or centrifugation (30,31). These detergent removal methods are quite often used for reconstitution of membrane proteins in liposomes.

Characterization of Liposomes

Size and homogeneity of liposomes are important in their use as drug carriers, since these characters affect the pharmacokinetics and distribution of liposomes in vivo. Light scattering, electron microscopy, NMR, and gel filtration are used for evaluating liposomal size and homogeneity (32,33)

Liposomes entrap hydrophobic materials in the lipid bilayer, and encapsulate hydrophilic materials in the aqueous inner phase. In the latter case, encapsulation depends on the volume of aqueous phase. This volume is called "trapped volume" and is designated in liters per mole of lipid. Trapped volume is usually measured by encapsulation of small aqueous markers. The encapsulation efficiency is a measure of how much material or drug is entrapped during the preparation of liposomes and is dependent on the preparation method, lipid concentration, and trapping material. The efficiency of trapping hydrophobic materials in the lipid bilayer is usually high. Liposomes formed by the freeze-thawing method also have a high trapping efficiency, e.g., nearly 90%, due to concentration of solute during freezing (34,35). In some experiments, an external material is entrapped in preformed liposomes. This is called active loading, and enables about 100% trapping of solutes (36,37)

Stability and Biodistribution of Liposomes

The fate of injected liposomes is drastically altered by administration route, dose and size, lipid composition, surface modification, and encapsulated drugs. Liposomes encapsulating drugs are often administered iv, therefore, the stability of liposomes in plasma is important. When liposomes composed of PC with unsaturated fatty acyl chains are incubated in the presence of serum, an efflux of internal solute from the liposomes is observed. This increase in permeability is caused by the transfer of phospholipids to high density lipoprotein (HDL) in serum (55). To reduce the efflux of liposomal contents, cholesterol is added as a liposomal component.

The half-life of liposomes administered in the blood stream is affected by the composition, size, charge, and fluidity. Liposomes with a small size or with a rigid lipid bilayer have a longer half-life (38,39). Large liposomes administered iv tend to accumulate at a lymph node near the injected site. This tendency can be useful for preventing metastases. Liposomes which pass through the lymph node have a tendency to accumulate in the RES, such as the liver and spleen (40,41). The disposition of liposomes is altered by the dose of liposomes as well as size or lipid composition of liposomes. Cholesterol rich liposomes are cleared slower due to their low uptake by RES (*42*). Since the RES uptake of liposomes is a saturatable process, the clearance rate of liposomal drugs decreases with increasing liposome doses or pre-dosing with empty liposomes (*43*)

Liposomes tend to remain at the injection site when they are administered intramuscularly or subcutaneously. Therefore, these administration routes are useful for slow and sustained release of drugs at the injection site.

Liposomes with a Specific Character

Many liposomes with a specific function, such as targeting, have been developed. Among them, four kinds of liposomes are described.

Temperature-Sensitive Liposomes. Liposomes that undergo phase transition at a specific temperature are used for site specific release of drugs, since the permeability of liposomal membrane increases drastically at the phase transition (Figure 2). When the target site is warmed, liposomes passing through the site are perturbed and release their contents. Dipalmitoyl PC (transition temperature at 41°C) is used as a main component (*24,44*). Temperature-sensitive liposomes have been further modified with a specific target site antibody for more effective drug release at the target site (*45*)

Target-Sensitive Liposomes. For targeting liposomes to a specific site, antibody-coated liposomes or ligand-coated liposomes are usually used. Target-sensitive liposomes composed of PE and palmitoyl-antibody have been developed; these are destabilized on binding to the target and release the contents at the specific site (*46*)

pH-Sensitive Liposomes. For cytoplasmic delivery of liposomal contents, fusogenic liposomes using virus fusogenic protein (*47*), or pH-sensitive liposomes have been developed (*48-50*). The latter can fuse with and/or destabilize biomembranes at low pH. When liposomes are internalized by a target cell, they are expected to fuse with or destabilize the endosomal membranes and to release their contents into the cytoplasm, since endosomes are known to be acidic. To endow low-pH fusogenic character, amphiphiles such as oleic acid are added to PE liposomes, or liposomes are modified with an imidazol-derivative. Cytosine arabinoside-encapsulating immunoliposomes composed of PE, oleic acid (8:2) and modified antibody are reported to be cytotoxic to target cells (*48*). This cytotoxicity is also reported to be inhibited by pretreatment of cells with weak bases such as chloroquine which raises the endosomal pH. Since oleic acid containing liposomes are not stable in plasma, other amphiphiles have been developed which endow fusogenic character to PE liposomes at acidic pH (*51*)

RES-Avoiding Liposomes. Since intravenously administered liposomes tend to accumulate in the RES, liposomal drugs work effectively when the target site is the RES. On the other hand, when the target site is not the RES, liposomal uptake by RES should be avoided. Liposomes that are not recognized by the RES have

Figure 2. Liposomes with a specific character a). temperature-sensitive liposome b). target-sensitive liposome c). pH-sensitive liposome. Closed triangles and rectangles in lipid bilayer indicate amphiphiles which change the liposome's hydration with pH changes.

prolonged circulation times. For this purpose, the liposomal surface has been modified with sialic acid (52). In fact, incorporation of GM_1 in liposomal membrane prevents liposomal uptake by macrophages (56). Glucuronic acid or poly(ethylene glycol) modified liposomes showed a similar tendency (53,54,57)

Literature Cited

1. *Liposome Technology*, Gregoriadis, G., Ed.; CRC Press: Boca Raton, FL, 1984.
2. *Macromolecules as Drugs and as Carriers for Biologically Active Materials*; Tirrell, D. A.; Donaruma, L. G.; Turek, A. B., Eds.; Ann. NY Acad. Sci., 1985; Vol. 446.
3. *Medical Application of Liposomes*; Yagi,, K., Ed.; Japan Scientific Soc. Press, Tokyo, Japan, 1986.
4. *Liposomes: Biophysics to Therapeutics*; Ostro, M. J., Ed., Marcel Dekker, Inc.: New York, NY, 1987.
5. *Liposomes as Drug Carriers*; Gregoriadis, G., Ed., John Wiley & Sons Ltd.: New York, NY, 1987.
6. *Drug Carrier Systems*; Roerdink, F. H.; Kroon, A. M., Eds., John Wiley & Sons Ltd.: New York, NY, 1989.
7. Hagiwarea, A.; Takahashi, T.; Oku, N. *CRC Crit. Rev. Oncology/Hematology*, **1989**, *9*, 319.
8. Mayer, L. D.; Hope, M. J.; Cullis, P. R. *Biochim. Biophys. Acta*, **1986**, *858*, 161.
9. Hub, H. H.; Zimmermann, U.; Ringsdorf, H. *FEBS Lett.*, **1982**, *140*, 254.
10. Oku, N.; Scheere, J. F.; MacDonald, R. C. *Biochim. Biophys. Acta*, **1982**, *692*, 384.
11. Szoka, F.; Papahadjopoulos, D. *Proc. Natl. Acad. Sci., USA*, **1978**, *75*, 4194.
12. Szoka, F.; Olson, F.; Heath, T.; Vail, W.; Mayhew, E.; Papahadjopoulos, D. *Biochim. Biophys. Acta*, **1980**, *601*, 559.
13. Gruner, S. M.; Lenk, R. P.; Janoff, A. S.; Ostro, M. J. *Biochemistry*, **1985**, *24*, 2833.
14. Weiner, A. L., Cannon, J. B., Tyle, P., In *Controlled Release of Drugs*; Rosoff, M., Ed.; VCH Publishers, Inc., New York, NY, 1989, p 217.
15. Adrian, G.; Huang, L., *Biochemistry*, **1979**, *18*, 5610.
16. Wong, M.; Anthony, F. H.; Tillack, T. W.; Thompson, T. E. *Biochemistry*, **1982**, *21*, 4126.
17. Szelei, J.; Duda, E. *Biochem. J.*, **1989**, *259*, 549.
18. Cullis, P. R.; de Kruijff, B. *Biochim. Biophys. Acta*, **1979**, *559*, 399.
19. Seddon, J. M.; Kay, R. D.; Marsh, D. *Biochim. Biophys. Acta*, **1983**, *734*, 347.
20. Verkleij, A. J. *Biochim. Biophys. Acta*, **1984**, *779*, 43.
21. Benita, S.; Poly, P. A.; Puisieux, F.; Delattre, J. *J. Pharm. Sci.*, **1984**, *73*, 1751.
22. Boggs, J. M. *Biochim. Biophys. Acta*, **1987**, *906*, 353.

23.   Oku, N.; Nojima, S.; Inoue, K. *Biochim. Biophys. Acta*, **1980**, *595*, 277.
24.   Weinstein, J. N.; Magin, R. L.; Cysyk, R. L.; Zaharko, D. S. *Cancer Res.*, **1980**, *40*, 1388.
25.   Decher, G.; Ringsdorf, H.; Venzmer, J.; Bitter-Suermann, D.; Weisgerber, C. *Biochim. Biophys. Acta*, **1990**, *1023*, 357.
26.   Hope, M. J.; Bally, M. B.; Webb, G.; Cullis, P. R. *Biochim. Biophys. Acta*, **1985**, *812*, 55.
27.   Mayhew, E.; Lazo, R.; Vail, W. J.; King, J.; Green, A. M. *Biochim. Biophys. Acta*, **1984**, *775*, 169.
28.   Pick, U. *Arch. Biochem. Biophys.*, **1981**, *212*, 186.
29.   Oku, N.; MacDonald, R. C. *Biochemistry*, **1983**, *22*, 855.
30.   Philippot, J.; Mutaftschiev, S.; Liautard, J.P. *Biochim. Biophys. Acta*, **1983**, *734*, 137.
31.   Alpes, H.; Allmann, K.; Plattner, H.; Reichert J.; Riek, R.; Schulz, S. *Biochim. Biophys. Acta*, **1986**, *862*, 294.
32.   Szoka, F.; Papahadjopoulos, D. *Annu. Rev. Biophys. Bioeng.*, **1980**, *9*, 467.
33.   Schurtenberger, P.; Hauser, H. *Biochim. Biophys. Acta*, **1984**, *778*, 470.
34.   Oku, N.; Kendall, D. A.; MacDonald, R. C. *Biochim. Biophys. Acta*, **1982**, *691*, 332.
35.   Mayer, L. D.; Hope, M. J.; Cullis, P. R.; Janoff, A. S. *Biochim. Biophys. Acta*, **1985**, *817*, 193.
36.   Mayer, L. D.; Bally, M. B.; Cullis, P. R. *Biochim. Biophys. Acta*, **1986**, *857*, 123.
37.   Mayer, L. D.; Tai, L. C. L.; Bally, M. B.; Mitilenes, G. N.; Ginsberg, R. S.; Cullis, P. R. *Biochim. Biophys. Acta*, **1990**, *1025*, 143.
38.   Senior, J.; Crawley, J. C. W.; Gregoriadis, G. *Biochim. Biophys. Acta,* , **1985**, *839*, 1.
39.   Senior, J.; Gregoriadis, G. *FEBS Lett.*, **1982**, *145*, 109.
40.   Rahman, A.; White, G.; More, N.; Schein, P. S. *Cancer Res.*, **1985**, *45*, 796.
41.   Lopez-Berestein, G.; Kasi, L.; Rosenblum, M. G.; Haynie, T.; Jahns, M.; Glenn, H.; Mehta, R.; Mavligit, G. M.; Hersh, E. M. *Cancer Res.*, **1984**, *44*, 375.
42.   Patel, H. M.; Tuzel, N. S.; Ryman, B. E. *Biochim. Biophys. Acta*, **1983**, *761*, 142.
43.   Ellens, H.; Morselt, H. W. M.; Dontje, B. H. J.; Kalicharan, D.; Hulstaert, C. E.; Scherphof, G. L. *Cancer Res.*, **1983**, *43*, 2927.
44.   Tomita, T.; Watanabe, M.; Takayuki, T.; Kumai, K.; Tadakuma, T.; Yasuda, T. *Biochim. Biophys. Acta*, **1989**, *978*, 185.
45.   Sullivan, S. M.; Huang, L. *Biochim. Biophys. Acta*, **1985**, *812*, 116.
46.   Ho, R. J. Y.; Rouse, B. T.; Huang, L. *Biochemistry*, **1986**, *25*, 5500.
47.   Martinez. O.; Kimura, J.; Henry, C.; Wofsy, L. *Exp. Cell Res.*, **1986**, *166*, 180.
48.   Conner, J.; Yatvin, M. B.; Huang, L. *Proc. Natl. Acad. Sci. USA*, **1984**, *81*, 1715.
49.   Connor, J.; Huang, L. *Cancer Res.*, **1986**, *46*, 3431.

50. Oku, N.; Shibamoto, S.; Ito, F.; Gondo, H.; Nango, M. *Biochemistry*, **1987**, *26*, 8145.
51. Collins, D.; Litzinger, D. C.; Huang. L. *Biochim. Biophys. Acta*, **1990**, *1025*, 234.
52. Gabizon, A.; Papahadjopoulos, D. *Proc. Natl. Acad. Sci. USA.*, **1988**, *85*, 6949.
53. Namba, Y.; Sakakibara, T.; Masada, M.; Ito, F.; Oku, N. *Chem. Pharm. Bull.*, **1990**, *38*, 1663.
54. Blume, G.; Cevc, G. *Biochim. Biophys. Acta*, **1990**, *1029*, 91.
55. Scherphof, G.; Morselt, H. *Biochem. J.*, **1984**, 221, 423.
56. Allen, T.M.; Chonn, A. *FEBS Lett.*, **1987**, 223, 42.
57. Klibanov, A.L.; Maruyama, K.; Torchilin, V.P.; Huang, L. *FEBS Lett.*, **1990**, 268, 235.

RECEIVED March 19, 1991

Chapter 4

Interactions between Polymeric Drug Delivery Systems and Biological Systems

Barriers and Opportunities

Kenneth J. Himmelstein

Himmelstein and Associates, 217 Gilbert Avenue, Suite B, Pearl River, NY 10965

The success of any drug delivery system depends upon how it interacts with the biological system to deliver its drug at the optimum rate and at the site where it is needed. This paper discusses some of the considerations that must be made in developing successful drug delivery systems and the opportunities available if a comprehensive assessment is made of the pharmacology, pharmacokinetics, and pharmacodynamics involved.

In general, the performance of any polymeric drug delivery system is highly dependent on both the therapeutic result desired and the milieu in which the system must operate. Both of these considerations are set by the interaction between the delivery system and the biological system in which it must operate. Most drug delivery systems, then, are intended to achieve one of the following three goals:

1. Take advantage of some property of a given route of administration to achieve a therapeutically optimum pattern of drug release.

2. Overcome one or more barriers posed by the available routes of administration to achieve a delivery pattern unachievable by conventional dosage forms.

3. Provide a specific interaction between the delivery and biological systems such as site-specific targeting, regional delivery, or localized delivery to improve the therapeutic index of a drug.

In this chapter an overview of both the opportunities and the problems presented by the biological system for the use of polymeric drug delivery systems will be presented. Since the area of biocompatibility of the delivery system is a well-known constraint also imposed by the biological system and is beyond the scope of this presentation, this (important) consideration will be ignored here. In order to examine how a delivery system interacts with the biological system to

0097–6156/91/0469–0034$06.00/0

achieve one or more of the above three goals, consider two major separate considerations: the route of administration and the ability to affect where internally the drug is to be transported.

Route of Administration

To date most drug delivery systems are designed to either overcome a barrier presented by or exploit an opportunity presented by a given route of administration. Often these two accomplishments are complementary. Each of the major routes present a unique set of barriers and exploitable characteristics. Consider the following major routes.

Oral. By far the most acceptable route of administration to the patient is to merely swallow the drug. yet even on the most superficial level, matters such as taste, odor, and size can negatively affect patient acceptance. The latter can have a negative impact on some delivery systems since low drug loadings are required for some polymeric systems to function properly. This logically leads to reservoir and other systems based on membrane transport for oral delivery. The barriers presented by the gastrointestinal tract are formidable. The general function of this system is to convert large molecules to small, innocuous, usable ones. Since some drugs are susceptible to degradation under certain conditions, e.g. acid hydrolysis, some dosage forms have been designed to pass through the upper tract to the more benign small intestine. Even more cleverly a few molecular systems have been designed to prevent absorption of drugs so as to deliver them to the colon, the site of action. Despite the decidedly hostile conditions presented by the alimentary canal, it represents an excellent route of administration for drug delivery systems since the transit time, especially for drugs well-absorbed throughout the system, can be rather long, often lasting up to sixteen or more hours. Thus a considerable effort has been made to develop systems for sustained or controlled release, based mainly on reservoir, osmotic, and matrix approaches. When delivery must be made to the upper tract only due to poor lower intestinal absorption, retention in the stomach can be manipulated by retaining the system with food, adhesion, or creating large masses that do not easily pass through the pylorus.

Parenteral. From a strict bioavailability viewpoint, there is no more efficient means of drug introduction that the parenteral route. The first pass effect associated with oral delivery is obviated, unstable drugs can be formulated immediately prior to administration, and high molecular weight drugs such as proteins and polypeptides can be administered. The route has a severe drawback however: injections are painful, can lead to infection, and in the case of repeated intravenous injection can lead to severe vascular problems. For these reasons polymeric delivery systems of both eroding and noneroding types have been considered. In addition to solving some of the issues noted, delivery systems which are injected or implanted can accomplish what no other system/route can: very long term dosing on the order of months or years. For chronic administration of potent drugs where compliance or constant dosing is important, this route is so unique that research efforts to use this route constitutes one of the largest areas of study.

Transmembranic Routes. In this area several different routes of administration are considered collectively. Among them are transdermal, buccal, nasal, rectal, and topical ophthalmic. All of these routes share many common problems and opportunities. The single largest common denominator is that in each case the drug must be absorbed across a biological membrane or structure by molecular diffusion. In all cases this is the single most important barrier since both the molecular weight of the drug and its physical properties (hydrophilicity/lipophilicity) can have a profound impact of the flux rates achievable. Indeed, instead of accepting the flux rate by passive diffusion as a given, significant effort has been exerted to influence delivery rates by such means as penetration enhancers and the use of other energy sources/potential gradients such as iontophoresis and ultrasound. Two other major issues are also associated with these routes: retention at the site of application and reproducible delivery. As an example, in topical ophthalmic drug delivery, typically less than five percent of an applied dose is locally bioavailable due to rapid removal of the drug from the applied sight. Thus, a major goal of drug delivery in these routes is to retain the drug, by means of a delivery system, at the desired site to maintain the flux of drug for sufficiently long periods of time to achieve improved absorption. Finally, because the rate of drug delivery depends not only on the flux but also the area of contact, often reproducibility issues must be addressed with respect to both the surface area involved and variable fluxes caused by thickness and transport property variations. Despite these barriers these routes of administration are of considerable interest. This is primarily due to two opportunities: these routes are noninvasive and avoid the degradation/metabolism common in the oral route. The desirability of the transdermal route, for example, cannot be denied. What could be easier than applying a plastic "bandage" to the skin to deliver a drug? The avoidance of the first pass effect and the acidic and proteolytic conditions of the stomach also is a powerful inducement. As a result, efforts in these areas of drug delivery research are certain to be of major importance.

Systemic Considerations.

Two basic questions arise from the biological system (excluding the route of administration discussed above) when the use of a polymeric drug delivery system is contemplated:

1. Will the biological system permit the achievement of the desired therapeutic goal?

2. Can the biological system be overcome with a delivery system to optimize therapeutic performance?

The latter question has been of major interest while the former has not been fully recognized.

The biological system is, of course, a complex set of processes that must be understood on micro-, meso-, and macroscopic levels before it can be determined a priori whether improved therapy can be achieved by the use of a polymeric drug delivery system. On the molecular-receptor level, the dynamics of the interaction are not often well understood. Thus the common assumption that constant concentration of a drug will achieve optimum therapy is simply not justified. As an example the use of transdermal nitroglycerin systems has enjoyed widespread use over the last decade. Yet, having the power to achieve nearly constant-rate delivery has demonstrated also that nitroglycerin tolerance leads to decreasing efficacy. As a result, it is not always possible to predict the desired release pattern based on existing known pharmacologic information even when the drug is currently used. On the mesoscopic scale, the properties of the biological system often dictate the success or failure of a delivery system. Consider the case of lipophilic drugs. Typically these compounds often have long terminal half-lives since they are easily taken up by adipose tissues which act as repositories for these drugs. Elimination then is controlled by the flow-rate of blood from these tissues. Thus the time scale of drug release may not significantly affect the overall distribution and elimination of these compounds. Finally, on the macroscopic level, systemic distribution and elimination can be manipulated just so much. Most target tissues to which one wishes to deliver the drug represent only a small fraction of the total volume (receptor, enzyme, tissue, or organ) available in the body. That means the systemic circulation has many more opportunities to deliver the drug to sites where untoward events can occur (metabolism, elimination, toxicity). Thus the strategy for the application of a given delivery system must be carefully weighed with respect to the entire biological system.

Notwithstanding the plaints above, the last question must be discussed: can a delivery system be used to overcome conditions imposed by the biological system (excluding those posed by the route of administration) to achieve optimized therapy? Two major approaches to this end have been proposed: regional drug delivery and targeted drug delivery.

Regional and local drug delivery is a rational approach when it is possible to readily identify the target tissue, e.g. a nondisseminated cancer. Here, it is possible to delivery the drug specifically to the target in high local concentration, yet avoiding the toxic consequences of systemic distribution. This is, as an example, the rationale for topical ophthalmic delivery: small doses of drug at high concentration can achieve the desired therapy without subjecting the entire system to toxic effects concommitant with systemic delivery. The obvious limitation with this approach is that these opportunities present themselves only infrequently.

Finally, the greatest goal of all from a physical delivery standpoint is whether or not it is possible to, with the aid of a polymeric delivery system, achieve site-specific as well as temporally optimum delivery of a drug. Polymeric carriers and vesicles have been investigated for such purposes. Whether or not they are feasible depends on three factors: the delivery, the biological system, and the drug itself. Most targeting strategies are based on either a specific affinity at the site of action for the delivery system, such as improved uptake of the system or specific release of the drug from the system at the site of action. That the system designer must recognize the mechanism to accomplish this is so obvious, it merits no comment.

One element that is often ignored in this evaluation though is the extent of the mechanism. For example, if enzymatic activity is to be used to release drug at the site of action, even if the activity/time/volume is one hundred times higher at the desired site than elsewhere (not unreasonable), but the volume of nonspecific activity is one thousand times as great (also not unreasonable), the system does not stand any chance of achieving site-specific delivery.

Finally, the interaction of the drug with the biological system has important consequences as to whether targeting approaches are feasible. Most drugs in use were selected from those which had intrinsic availability to the site of action by virtue of the discovery method: try it in the intact animal. These drugs naturally have demonstrated access to the target site. Therefore when they are released from the delivery system they are also quite likely to leave the site of action rapidly. This suggests hope for the future. As drugs are designed by rational means, are the products of biotechnology, or are screened in in vitro culture, they will not have intrinsic access to the site of action a priori and may actually require site-specific delivery systems.

Conclusions

In this brief overview some considerations with respect to the biological system vis-a-vis polymeric drug delivery have been qualitatively presented. Delivery systems can be designed to either overcome a barrier or take advantage of a property associated with a route of administration. Once introduced into a biological system, a delivery system's operation must be consistent with the operation of the biological system to achieve the desired goal based on temporal release pattern.

Finally, site-specific delivery also depends on the extent of competing processes in the biological system as well as the interaction of the drug itself with the biological system. Therefore successful design of polymeric drug delivery approaches requires a comprehensive appreciation of the pharmacology, pharmacokinetics, and pharmacodynamics involved.

Bibliography

For further general reading in the area of drug delivery and biological systems interaction, the following texts and papers are recommended as a starting place.

Li, V.H.K.; Lee, V.H.L.; Robinson, J. R.; "Influence of drug properties and routes of drug administration on the design of sustained and controlled release systems"; In *Controlled Drug Delivery;* Robinson, J.R., Lee, V.H.L., Eds.; Drugs and the Pharmaceutical Sciences: M. Dekker, New York, NY, 1987, Vol. 29; pp. 9-94.

Silber, B.M.; Bialer, M., Yacobi, A.; "Pharmacokinetic/pharmacodynamic basis of controlled drug delivery"; In *Controlled Drug Delivery;* Robinson, J.R., Lee, V.H.L., Eds.; Drugs and Pharmaceutical Sciences: M. Dekker, New York, NY, 1987, Vol. 29, pp. 213-252.

Welling, P.G.; Deobrinska, M.R.; "Dosing considerations and bioavailability assessment of controlled drug delivery systems"; In *Controlled Drug Delivery;* Robinson, J.R., Lee, V.H.L., Eds.; Drugs and the Pharmaceutical Sciences: M. Dekker, New York, NY, 1987, Vol. 29, pp. 253-292.

Stella, V.J.; Himmelstein, K.J.; "Prodrugs and site-specific delivery"; *J. Med. Chem.,* **1981**, 23, 1275.

Lee, V.H.L.; Robinson, J.R.; "Review: topical ocular drug delivery: recent developments and future challenges"; *Journal of Ocular Pharmacology,* **1986**, 1, 67.

Mayersohn, M.; "Principles of drug absorption"; In *Modern Pharmaceutics;* Banker, G.S., Rhodes, C.T., Eds.; Drugs and the Pharmaceutical Sciences: M. Dekker, New York, NY, 1990, Vol. 40, pp. 23-90.

Benet, L.Z.; "The effect of drug administration and distribution on drug action"; In *Modern Pharmaceutics;* Banker, G.S., and Rhodes, C.T., Eds.; Drugs and the Pharmaceutical Sciences: M. Dekker, New York, NY, 1990, Vol. 40, pp. 181-204.

Tomlinson, E.; "Site specific drug delivery using multiparticulate systems", In *Modern Pharmaceutics,* Banker, G. S., Rhodes, C. T., Eds.; Drugs and the Pharmaceutical Sciences: M. Dekker, New York, NY, 1990, Vol. 40, pp. 673-694.

RECEIVED March 19, 1991

POLYMERIC DRUGS
AND DRUG CONJUGATES

Polymeric Drugs and Drug Conjugates

Although the majority of drugs used during the past century for treatment of human diseases have all been small molecules, it has become increasing evident during the last ten years that polymers or macromolecules may have important therapeutic benefits. Many of the functions within the body are regulated by means of macromolecules and with the advent of genetic research during the past decades, many of these macromolecules classified as peptides and proteins have been produced in commercial quantities and a number have been approved for use in humans. However, there are polymers other than the poly(amino acids) which have biological activity. Some of these are natural body constituents and others are synthetic polymers.

In addition to being used alone for therapeutic effects, polymers can also be used as polymer-drug conjugates. In these instances, a known drug is chemically attached to the polymer. Depending upon the method of attachment, the drug can be effective while still attached to the polymer or it has to be released from the polymer within the body by some type of reaction before it can exert its biological response. The papers in this section (Chapters 5-13) present the results of some recent developments in the area of polymeric drugs and polymer-drug conjugates.

One group of polymers that has not received the attention given to the peptides and proteins are the polysaccharides, and the first four papers in this section (Chapters 5-8) are devoted to this important class of polymers. In Chapter 5, the properties of a genetically engineered, yeast derived glucan polysaccharide that activates the immune system is described. In Chapter 6, a similar type of glucan polysaccharide that can be used as a vaccine adjuvant or a drug delivery system is presented. The drug is either physically entrapped in microcapsules of the polysaccharide or covalently attached to the polymer. Thus, we have a polymer that is active itself and can be conjugated to another active drug. Chapter 7 discusses the covalent attachment of a compound to another polysaccharide, heparin. In this case, the polysaccharide is a well-known drug whereas the compound attached to it is not a drug but rather a fluorescein label used to study the drug in vitro and in vivo. In Chapter 8, polysaccharides with no reported biological activities are conjugated to 5-fluorouracil, a well-known antitumor drug. These four chapters demonstrate the versatility of polymeric materials for use as either drugs or conjugate carriers for other drugs.

The next four chapters (Chapters 9-12) deal with a variety of polymer-drug conjugates. Chapter 9 shows the advantages of conjugating a biologically active water-soluble polymer to methotrexate, a well-known antitumor agent. Chapter 10 discusses the preparation of other water-soluble polymer-drug conjugates. In this case, polyethylene glycol was linked to proteins through stable urethane linkages.

The next two chapters describe polymer-drug conjugates in which the polymers are biodegradable polymers. In Chapter 11, the synthesis and characterization of poly(α-amino acid) conjugates with two drugs is presented as well as the in vitro and in vivo release of the drugs when implanted subcutaneously. Chapter 12 shows the effect of a biodegradable and nonbiodegradable polymeric carrier upon release of a conjugated drug during the transcellular transport process. This release at the cellular level may determine the efficacy of the polymer-drug conjugate in effecting a biological response.

The last chapter of this section (Chapter 13) also deals with a polymer conjugate. However, in this case, the polymer is a synthetic polypeptide covalently bound to a fluorophore and the main emphasis in this paper is to evaluate the effect of polymer molecular weight upon rate of absorption of the polypeptide in the lung, a totally new method of drug delivery.

Chapter 5

PGG-Glucans

A Novel Class of Macrophage-Activating Immunomodulators

S. Jamas[1], D. D. Easson, Jr.[1], G. R. Ostroff[1], and A. B. Onderdonk[2]

[1]Alpha-Beta Technology, Inc., Two Biotech Park, Worcester, MA 01605
[2]School of Veterinary Medicine, Tufts University, Boston, MA 02130

PGG represents a novel class of genetically engineered, yeast derived, carbohydrate activators of early non-specific host defenses. The primary structure of PGG has been optimized for enhanced macrophage stimulating properties via β-glucan receptors found on monocytes and neutrophils. Compositional and structural analyses have shown that PGG is a β(1-3)/β(1-6) linked glucose polymer with a unique branching structure. The altered primary structure of the PGG polymer affects its spacial conformation which in turn increases its avidity for the β-glucan receptor as compared to naturally occurring β-glucans. Activation of monocytes and neutrophils through the β-glucan receptor results in the balanced and rapid amplification of host immune defenses involving a cascade of interactions mediated by the release of monocyte-derived products. PGG treated animals show increased levels of monocyte and neutrophil phagocytic activity and increased protection against a wide variety of pathogenic challenges. An intravenous (iv) injection of 20 µg/animal of soluble PGG (Betafectin, Alpha-Beta Technology, Inc.) 4-5 hours prior to challenge with 1×10^7 CFU/animal of *E. coli* or *S. aureus* significantly reduced mortality as compared to saline treated control groups. These results suggest that PGG, produced as a soluble, highly purified, nonpyrogenic and nonantigenic carbohydrate solution for intravenous/intramuscular (iv/im) administration maybe useful for the prevention and treatment of infectious diseases.

β-Glucan polysaccharides commonly present in the cell walls of yeasts and fungi have been shown to possess a range of immunomodulating properties which include antitumor (*1,2,3*), anti-infective (*3,4*), and wound healing (*5,6*) activities. It is not clear if all the different β-glucan polymers act through a common mechanism of action, however, human monocytes and neutrophils are known to possess cell surface receptors which are ligand specific for the β-D-(1-3)-linked glucopyranosyl glucans (7). In fact, neither mannans, galactans nor the α-anomeric and β(1-4)-linked glucose polymers have any avidity for this receptor (8).

Consistent with the above observations, the immunoactive glucans such as curdlan (9), lentinan (10), scleroglucan (11), schizophyllan (12) and yeast glucan (13) share a common β-D(1-3)-linked glucopyronosyl backbone. Some of these polymers also contain β-D(1-6)-linked glucopyranosyl branches through the 3,6-di-O-substituted C-6 atom of the backbone residues.

This paper describes the structural and biological characterization of a genetically engineered β-glucan derived from yeast. This glucan, known as PGG glucan, has a distinct primary structure as characterized by ^{13}C-NMR spectroscopy, resulting in a significantly increased affinity for the β-glucan receptor present on peripheral monocytes and neutrophils as compared to wild-type yeast glucans (14). Immunoactivation by PGG glucan via the β-glucan receptor results in the enhancement of monocyte and neutrophil phagocytosis and microbicidal capacity and enhances the production of cytokines, colony stimulating factors and inflammatory factors. In vivo PGG glucan (Betafectin) shows significant protective activity in preclinical acute sepsis models against a wide variety of bacterial and fungal pathogenic challenges.

EXPERIMENTAL

Materials. Soluble PGG Glucan (Betafectin, Alpha-Beta Technology, Worcester, MA) was prepared from *Saccharomyces cerevisae* strain R4 cells (14).

Structural Characterization. ^{13}C-NMR Spectra of PGG glucan preparations (15 mg/mL in 0.5 M NaOD) were recorded with a Bruker Model AC 200 at 50.3 MHz and all chemical shifts were expressed in parts per million downfield from an internal tetramethylsilane (TMS) standard.

Conformational Characterization. Glucan solutions (5 mg/mL) and congo-red (44 µM) were prepared in a range of NaOH concentrations (0.01-0.15 M). An equal volume of glucan and congo-red were mixed at each NaOH concentration and the absorbance maximum of their visible absorption spectrum (450-650 nm) was determined using a Beckman model DU-70 spectrophotometer. Each measurement was referenced against 22 µM congo-red at the respective NaOH concentration.

Characterization of Anti-Infective Biological Activity. Acute sepsis models utilizing either *Escherichia coli or Staphylococcus aureus* intraperitoneal challenge were developed to evaluate the anti-infective properties of PGG in mice.

Male CD-1 mice were allowed 3 days to stabilize after arrival at the test facility. Groups of 10 mice each received 0.2 mL of the various concentrations of PGG in sterile saline by bolus intravenous (iv) injection (transthoracic cardiac puncture). A control group received 0.3 mL of sterile saline. Mice were returned to their cages, maintained on food and water ad libitum and were challenged 3-4 hours after administration of PGG by intraperitoneal injection with either 0.1 mL (1×10^8 CFU/mL) of an *E. coli* culture or 0.1 mL (1×10^8 CFU/mL) of *S. aureus*. Survival was recorded at 48 hours after challenge.

RESULTS

Structural Characterization. A common technique in determining linkage type and structure in glucans is ^{13}C-NMR spectroscopy. The number and relative intensities of ^{13}C-signals in a given spectrum can be used to determine linkage configurations and positions in glucan polymers. Figure 1 presents the ^{13}C-NMR spectrum of PGG glucan. Table I summarizes the structural information obtained for the engineered PGG glucan and compares this to the structures of other immunoactive glucans which have been documented in the literature.

The engineered PGG glucan exhibited two distinct structural differences in its primary structure when compared to the other immunoactive glucans. Firstly, the PGG glucan contains the highest branching frequency which was estimated at approximately one branch point per two glucose residues in the backbone. Secondly, a unique chemical shift at 69 ppm in the ^{13}C-spectrum of PGG indicates the presence of a 6-0-substituted C-6 which has been interpreted as that of internal β-D(1-6)-linked moieties in the branch. Consequently unlike the other listed glucans, PGG glucan contains branches of greater than one glucose unit in length.

Conformational Behavior of Glucans. Glucans that contain β-D(1-3)-linked glucopyranosyl backbones have been shown to form triple helical conformations in solution (*9,15*). These ordered states are the result of steric limitations imposed by the glycosidic linkages and the stabilizing contribution of cooperative H-bonding (or other electrostatic interactions). The triple helical conformation of glucans will also persist to varying degrees depending on the frequency of β-D(1-6)-linked branching. The effect of the increased β-D(1-6) branching on the conformation of PGG was investigated using a congo-red binding assay. The congo-red dye forms a complex with triple helical glucans which has a distinct absorbance maximum (*10,16*). Thus it is possible to detect a conformational transition from triple helix to single helix (or a random coil conformation) by measuring the absorbance maxima of glucan-congo red solutions under different conditions.

Figure 2 compares the conformational transition curves of wild-type yeast glucan (branch frequency = 0.20) and PGG (branch frequency = 0.50). Wild-type yeast glucan required approximately 0.1M NaOH to disrupt the triple helical conformation, whereas this transition is observed at approximately 0.04 M NaOH with PGG. This trend is consistent with the observation that curdlan, an entirely linear β-D(1-3)-linked glucan, requires approximately 0.25M NaOH to disrupt the ordered conformation (*16*) . Hence, it is concluded that the highly branched PGG molecules only form weak inter-chain associations resulting in the formation of predominantly single-helical zones.

Anti-Infective Activity of PGG. Acute sepsis models utilizing either *Escherichia coli* or *Staphylococcus aureus* intraperitoneal challenge were developed to evaluate

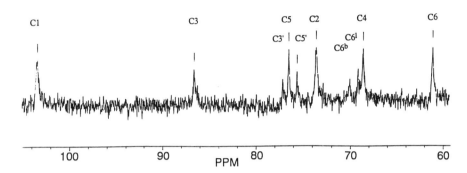

Figure 1: ^{13}C-NMR spectrum of PGG Glucan. Carbon assignments C1-C6 represent β(1-3)linked glucose units, C3' and C-5' represent the resolved C3 and C5 of the β(1-6) branch. C-6^b represents the β(1-6) branch point on a β(1-3)linked glucose residue and C-6^1 represents the internal β(1-6)linked glucose residue of the β(1-6) branch.

Table I. Branching and Linkage Structure of Glucans

Glucan	Branching Frequency[1]	Branch Length
Curdlan (16)	0	0
Soluble Phosphorylated Glucan (3)	0	0
Alkali Insoluble Yeast Glucan (13)	0.03	1
Scleroglucan (13)	0.33	1
Schizophyllan (12)	0.33	1
Lentinan (10)	0.40	1
PGG	0.50	2-4

[1] Branching Frequency - number of branches/number of glucose moieties per repeating unit.

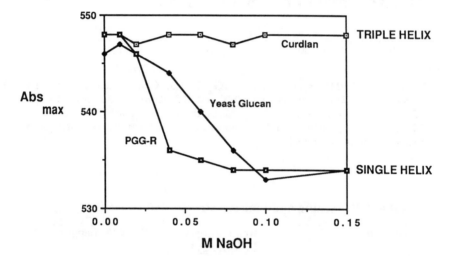

Figure 2: Effect of branch frequency on glucan conformation. Conformational characterization of glucans was carried out as described in the experimental section. Curdlan is a linear β(1-3)linked glucan; Yeast glucan has a 30% β(1-6) branch frequency and PGG-R glucan has a 50% β(1-6) branch frequency. The Congo Red-single/triple helix complex absorption maxima are indicated.

the anti-infective activity of PGG in mice. Figures 3 and 4 summarize the typical dose-responses observed with PGG glucan in these models. A single dose of PGG glucan (0.02-0.05 mg/kg) administered 3-4 hours before challenge effected significant reductions in mortality from intraperitoneal sepsis. At high PGG glucan doses (>2 mg/kg) the protective effects were lost, an unexplained characteristic of immunomodulators.

DISCUSSION

To date, polysaccharides have been commercialized in applications which primarily exploit their physical and rheological properties. β-Glucans are unique examples of polysaccharides that also have specific biological properties which involve interactions between the polysaccharide and cell surface proteins (receptors). Due to primary structural heterogeneity amongst various β-glucans it can be postulated that they can exist in a range of conformations which, much like proteins, affect their biological activity.

This report describes the effect of structure and conformation on the immunomodulatory activity of a yeast β-glucan. Specifically, an engineered β-glucan, PGG glucan, was shown by ^{13}C-NMR analysis to contain significantly higher β-D(1-6)-branching when compared to other known naturally occurring β-glucans. In comparison to natural yeast glucan, PGG glucan β-D(1-6)-branch frequency was increased from 30% to 50% and β-D(1-6)-branch length was increased from 1 to 2-4 glucose residues per repeat unit.

This increase in β-D(1-6)-branching affected the ability of PGG glucan to exist in solution as a highly ordered triple-helical conformation, as evidenced by intercalating dye analysis (*10, 16*). PGG glucan exists in a solution conformation composed largely of less ordered single helical regions. Other naturally occurring glucans containing lower β-D(1-6)-branching, such as curdulan, lentinan, scleroglucan, schizophyllan and wild-type yeast glucan, exist in an ordered triple helical solution conformation (*10, 16, 17, 18*). As a consequence of this conformational difference, PGG glucan has shown a 35 fold higher affinity for the β-glucan receptor of human monocytes when compared to wild-type yeast β-glucan (*14*). Thus PGG glucan is a very avid ligand for the β-glucan receptor.

The immunomodulating properties of naturally occuring glucans has been widely studied and shown to include antitumor, (*1,2,3*) anti-infective (*3, 4*) and wound healing activities (*5,6*). These therapeutic activities appear to be mediated via the β-glucan receptor present on moncytes and neutrophils specific for the β-D(1-3)-linked glucopyranosyl glucans (*7*). Activation of moncytes by β-D(1-3)-linked glucopyranosyl glucans results in the coordinated stimulation of phagocytosis (*7,8*), and enhanced cytokine (*19*), inflammatory response factor (*20*), and colony stimulating factor production (*21*). PGG glucan (Betafectin), a ligand engineered for high β-glucan receptor avidity, has recently been shown to effectively activate moncytes and neutrophils resulting in enhanced phagocytosis and cytokine, inflammatory factor and colony stimulating factor induction (*22*).

The anti-infective therapeutic potential of PGG glucan (Betafectin) was demonstrated by its ability to protect mice from lethal *E. coli* and *S. aureus*

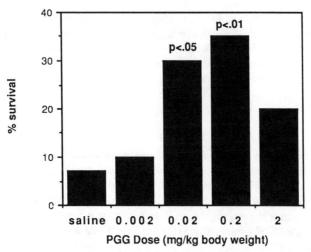

Figure 3: PGG Dose-response in an *E. coli* Peritoneal Sepsis Challenge. Groups of 10 mice each received 0.2 mL of the various concentrations of PGG in sterile saline by bolus intravenous (iv) injection (transthoracic cardiac puncture). A control group received 0.3 mL of sterile saline. Mice were challenged 3-4 hours after administration of PGG by intraperitoneal injection with 0.1 mL (1×10^8 CFU/ml) of an *E. coli* culture. Survival was recorded at 48 hours after challenge.

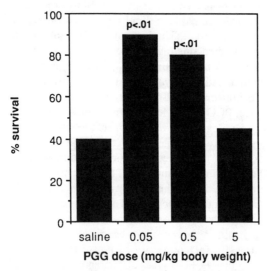

Figure 4: PGG Dose-response in an *S. aureus* Peritoneal Sepsis Challenge. Groups of 10 mice each received 0.2 mL of the various concentrations of PGG in sterile saline by bolus intravenous (iv) injection (transthoracic cardiac puncture). A control group received 0.3 mL of sterile saline. Mice were challenged 3-4 hours after administration of PGG by intraperitoneal injection with 0.1 mL (1×10^8 CFU/mL) of *S. aureus*. Survival was recorded at 48 hours after challenge.

infections. A single iv injection of 0.02-0.2 mg/kg body weight was effective at significantly reducing mortality in these acute sepsis models. Similar studies have demonstrated PGG glucan induced protection against a wide variety of other bacterial and fungal pathogenic challenges. These results suggest that PGG glucan (Betafectin) produced as a soluble, highly purified, nonpyrogenic and nonantigenic polysaccharide solution for iv/im administration may be useful for the prevention and treatment of infectious diseases.

Literature Cited

1. Zakany, M.; Chihara, G.; Fachet, J. *Int. J. Cancer 1980, 25*, 371.
2. Singh, P.; Whistler, R. L.; Tekuzan, R.; Nakahara, W. *Carb. Res.* **1974**, *37*, 245.
3. DiLuzio, N. R.; Williams, D. L.; McNamee, R. B.; Edwards, B. F.; Kitahama, A. *Int. J. Cancer* 1979, *24*, 773.
4. Reynolds, J. A.; Kastello, M. D.; Harrington, D. G.; Crabbs, C. L.; Peters, C. J.; Jemski, J. V.; Scott, G. H.; DiLuzio, N. R. *Infec. Immun.* **1980**, *30*, 51.
5. Leibevich, S. J.; Danon, D. *J. Reticul. Soc.* **1980**, *27*, 1.
6. Browder, W.; Williams, D. L.; Lucore, P.; Pretus, H.; Jones, E.; McNamee, R. B. *Surgery* **1988**, *104*, 224.
7. Czop, J. K.; Austen, F. K. *J. Immunol.* **1985**, *134*, 2588.
8. Czop,J. K. *Adv. Immunol.* **1986**, *38*, 361.
9. Saito, H.; Ohki, T.; Sasaki, T. *Biochemistry* **1977**, *16*, 908.
10. Saito, H.; Ohki, T.; Takasuka, N.; Sasaki, T. *Carbohydr. Res.* **1977**, *58*, 293.
11. Rinaudo, M.; Vincendon, M. *Carbohydr. Pol.* **1982**, *2*, 135.
12. Yanaki, T.; Ito, W.; Tabata, K.; Kojima, T.; Norisuye, T.; Takano, N.; Fujita, H. *Bioph. Chem.* **1983**, *17*, 337.
13. Manners, D. J.; Masson, A. J.; Patterson, J. C. *Biochem. J.* **1973**, *135*, 19.
14. G. R. Ostroff, D.D. Easson Jr., and S. Jamas. 198th ACS National Meeting. September 1989.
15. Bluhm, T. L.; Sarko, A. *Can. J. Chem.* **1977**, *55*, 293.
16. Ogawa, K.; Tsurugi, J.; Watanabe, T. *Chem. Lett.* **1972**,*689*, 692.
17. Norisuye, T.; Yanaki, T.,; Fujita, H. *J. Polym. Sci. Polym. Phy. Ed.* **1981**, *18*, 547.
18. Bluhm, T.; Deslandes, Y.; Marchessault, R. *Carbohydr. Res.* **1982**, *100*, 117.
19. Sherwood, E. R.; Williams, D. L.; McNamee, R. B.; Jones, E. L.; Browder, I. W.; Di Luzio, N. R. *Int. J. Immunpharmc.* **1987**, *9*, 261.
20. Czop, J.K.; Austen, F.K. *Proc. Natl. Acad. Sci. USA.* **1985**, *82*, 2751.
21. Patchen, M. L.; Macvittie, T. J. *J. Immunopharm.* **1986**, *8*, 407.
22. Symposium: The Beta-Glucan Receptor and Response to PGG, International Congress for Infectious Diseases, Montreal, Canada, July 1990.

RECEIVED March 19, 1991

Chapter 6

A New β-Glucan-Based Macrophage-Targeted Adjuvant

G. R. Ostroff, D. D. Easson, Jr., and S. Jamas

Alpha-Beta Technology, Inc., Two Biotech Park, Worcester, MA 01605

Through polysaccharide engineering, the composition, structure and porosity of novel glucan carbohydrate microcapsules (Adjuvax) can be controlled to yield sophisticated targeted antigen or drug delivery vehicles. The unique β(1-3)/β(1-6) - linked glucan structure of Adjuvax allows the targeting of antigens to macrophages or neutrophils through interactions with the β-glucan receptor found uniquely on the surface of these cell types. Adjuvax- ligand complexes are formed by physically entrapping or crosslinking the ligand within the β-glucan microcapsule. Sustained release rate is dependent upon microcapsule porosity, ligand molecular weight, degree of Adjuvax - ligand crosslinking, and the rate of carbohydrate biodegradation. Microcapsule porosity is controlled by varying the ratio of the β(1-3)/β(1-6) linkages in the carbohydrate molecule through genetic and process manipulations. The in vitro sustained release rate of entrapped proteins ranged from 1 hour for a 12 kD protein to 6 hours for a 150 kD protein. Adjuvax - ligand crosslinking increased the sustained release rate of the 12 kD protein to greater than 24 hours. Adjuvax coadministered with bovine serum albumin or crosslinked to a series of peptide antigens increased mouse antibody titers, 1,000 fold over antigen only controls. Adjuvax stimulation of antibody titer was equivalent to Complete Freund's Adjuvant (CFA) without the toxicity and histopathology associated with use of CFA. These results suggest that the use of this safe, non-antigenic, newly defined carbohydrate adjuvant may prove useful in new generation vaccines.

Adjuvax is a non-antigenic, polysaccharide vaccine adjuvant and drug delivery vehicle designed to deliver a broad spectrum of antigens or drugs to the macrophage cell surface. The active component of Adjuvax is a branched β-glucan polymer of

0097–6156/91/0469–0052$06.00/0
© 1991 American Chemical Society

defined structure, known as PGG glucan, which has been engineered for optimal biologic activity (*1*). The PGG glucan molecule is formulated as a highly purified carbohydrate microcapsule combining three key adjuvant properties:

- macrophage targeting of antigens
- macrophage activation
- sustained release of antigens

The PGG glucan molecules bind to specific β-glucan receptors on macrophages and neutrophilic phagocytes (*2,3,4,*) and activate a cascade of immunologic responses (*2,5,6,7,*) which promote the production of higher titers of serum antibodies to specific antigens (*8,9,10*). This report demonstrates the feasibility for developing Adjuvax as a vaccine adjuvant.

EXPERIMENTAL

Materials. Microspherical PGG glucan (Adjuvax, Alpha-Beta Technology, Worcester, MA) was prepared from *Saccharomyces cereviseae* strain R4 cells (*11*). Zymosan, cytochrome c (cyt c), bovine serum albumin (BSA), yeast alcohol dehydrogenase (ADH), Complete Freunds Adjuvant (CFA) and Incomplete Freunds Adjuvant (IFA) were purchased from Sigma Chemical Co. (St. Louis, MO).

Measurement of Adjuvax Phagocytosis. Adjuvax microcapsules were assayed at three doses for their relative capacity to trigger phagocytosis after 30 minutes of exposure to adherent human monocytes (*2*). Adjuvax concentrations of 5×10^6/mL to 6×10^7/mL corresponding to particle-to-cell ratios of approximately 5 to 50 were used. The number of Adjuvax particles ingested by at least 300 monocytes was determined by direct visual observation with a 1000X light microscope. The results are expressed as the percentage of monocytes ingesting ≥ 3 particles.

Sustained Release Studies. The ligand proteins BSA, cyt c, and ADH dissolved at 10 mg/mL in water, were loaded into Adjuvax microcapsules by hydration (0.07 mL protein/mg Adjuvax) for four hours at room temperature. A control, run in parallel contained no Adjuvax. The loaded microcapsules were dried by lyophilization and ground to a fine powder. Protein release was monitored by hydrating the powder with saline and agitating at 37°C. Samples were removed at regular intervals, centrifuged to remove the Adjuvax microcapsules, and soluble supernatants containing the released protein were assayed (absorbance, 280 nm). Percent protein retained was calculated by dividing the absorbance (280 nm) of each time point by the absorbance (280 nm) of the protein only control.

Adjuvax - Ligand Crosslinking. Whole glucan particles containing crosslinked cyt c were prepared by first reacting 5 mg cyt c with 2.5 mg of the heterobifunctional crosslinking reagent sulfosuccinimidyl 6- (4'-azido-2'-nitrophenylamino-hexanoate (sulfoSANPAH, Pierce Chemical Co., Rockford, IL) in 1 mL of 10 mM sodium phosphate buffer pH 7.4 for 16 hours at 25°C in the dark. One mL of the sulfoSANPAH-cyt c conjugate was swelled into the PGG glucan microcapsule by mixing with 150 mg of dry Adjuvax, followed by incubation at 25C for 2 hours in the dark. The sulfoSANPAH-cyt c conjugate was crosslinked to the Adjuvax by exposure to bright light. The unreacted sulfoSANPAH, cyt c and sulfoSANPAH-cyt c conjugate were removed by washing with saline. The crosslinked Adjuvax sulfoSANPAH-cyt c conjugate was dried by lyophilization and stored at 4°C.

In vivo Adjuvant Characterization of Adjuvax. The adjuvant effect of Adjuvax in mice was determined by measuring the antibody production to BSA and a 55 amino acid (P55) antigen coupled to BSA in two different Adjuvax formulations. One formulation was prepared by the physical entrapment of the antigen within the Adjuvax microcapsules. The second formulation contained the antigen crosslinked to the Adjuvax microcapsules. A standard adjuvant control contained antigen mixed with CFA (primary immunization) or IFA (secondary immunization).

For each antigen, three groups of ten mice were immunized subcutaneously on day 1 with 0.2 mL of the entrapped formulation, 0.2 mL of the crosslinked formulation or 0.2 mL of a CFA- antigen emulsion. A secondary immunization was administered on day 14 with 0.2 mL of the same Adjuvax formulations or 0.2 mL of an IFA-antigen emulsion. On day 27, serum was collected and analyzed for anti-P55 antibody or anti-BSA antibody by ELISA.

RESULTS

Adjuvax Activation of Phagocytosis by Human Monocytes. The PGG carbohydrate structure of Adjuvax was optimized using in vitro assays for its ability to interact with the β-glucan receptor of human monocytes and neutrophils. This receptor is identified with the phagocytic receptor for particulate activators of the alternate complement pathway (2). Figure 1 compares the engineered Adjuvax microcapsule to zymosan, the typical particulate activator of phagocytosis. These results demonstrate the increased potency of Adjuvax over standard glucans (zymosan) to be targeted to, bound and phagocytosed by macrophages.

Sustained Antigen and Drug Release from Adjuvax. Physical entrapment of certain ligands (antigen or drug) within the hollow Adjuvax microcapsules represents the most basic delivery mode. The ligand can be trapped within the microcapsules by a variety of techniques tailored specifically to the physical properties of the molecule. The sustained release of ligand from Adjuvax microcapsules under physiologic conditions is monitored in vitro and is dependent upon natural diffusion. In vivo release is additionally dependent on biodegradation of the Adjuvax PGG matrix. Both these mechanisms are controlled by the primary structure of the engineered PGG molecules and are a function of the molecular weight of the entrapped ligand.

The release rate of proteins that have been physically entrapped within Adjuvax microcapsules can be used as a preliminary indication of the sustained release potential of this system. Three proteins of different molecular weight (cyt c = 14 kD; BSA = 67kD and ADH = 150 kD) were entrapped within Adjuvax microcapsules. Figure 2 shows the effect of antigen molecular weight on sustained release rate. The time to release 50% of the protein (T_{50}) from the Adjuvax microcapsules increased as the molecular weight of the antigen increased (31 minutes for cyt c; 80 minutes for BSA; and 200 minutes for ADH).

These sustained release results show that entrapped low molecular weight ligands are rapidly released from Adjuvax microcapsules. To extend the application of Adjuvax microcapsules to low molecular weight molecules, chemical attachment techniques have been developed which do not interfere with microcapsule targeting or ligand functionality. Proteins and peptides were covalently crosslinked to Adjuvax microcapsules through amino groups using the heterobifunctional crosslinking agent sulfosuccinimidyl 6-(4'-azido-2'-nitrophenylamine) hexonate. Figure 3 compares the release rates of ligand from a crosslinked Adjuvax x cyt-c conjugate and a physically entrapped Adjuvax:cyt-c formulation. Crosslinking significantly increased the retention time of cyt-c from 31 minutes to greater than 144 hours. Release of crosslinked ligands was dependent upon ligand or polysaccharide degradation.

In Vivo Adjuvant Effect of Adjuvax in Mice. Adjuvax was compared to CFA for its ability to stimulate antibody production to protein (BSA) and peptide (a 55 amino acid residue oligopeptide) antigens in two different Adjuvax formulations. One formulation was prepared by physical entrapment of the antigens within the Adjuvax microcapsules, the second formulation contained the antigens crosslinked to the Adjuvax microcapsules. The BSA formulation contained an Adjuvax/antigen ratio of 100 mg Adjuvax/10 mg antigen per dose (0.2mL), the peptide formulation contained an Adjuvax/antigen ratio of 100 mg Adjuvax/50 mg antigen per dose (0.2mL). Controls containing the same amounts of antigen were mixed with CFA, a standard adjuvant. Figures 4 and 5 summarize the antibody titer results.

These results show that the covalently crosslinked Adjuvax formulations were superior to the physically entrapped Adjuvax formulations. In addition, the Adjuvax crosslinked formulations were as effective in stimulating antibody titers as Freund's Adjuvant. Furthermore, animals immunized with Adjuvax did not experience local inflammatory reactions or granuloma formation which was observed with all CFA/IFA immunized animals.

DISCUSSION

We have established that Adjuvax is effectively targeted to the macrophage via the β-glucan receptor. Adjuvax has a novel, engineered glucan structure (PGG glucan) that has a higher avidity for the β-glucan receptor and is thus more potent than previousl described glucans. In addition, Adjuvax has been recently shown to effectively activate monocytes and neutrophils resulting in enhanced cytokine (IL-1α, IL-1β, IL-6 and TNF) and colony stimulating factor production (GM-CSF) (*12*). Thus, the strong adjuvant properties of Adjuvax can be attributed to the unique targeted antigen delivery and macrophage activation features of the PGG glucan matrix.

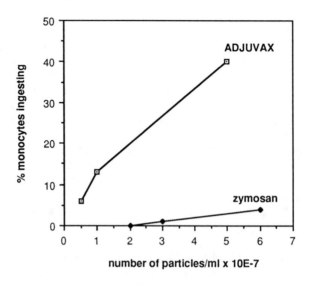

Figure 1. Stimulation of phagocytic capacity in human monocytes with Adjuvax.

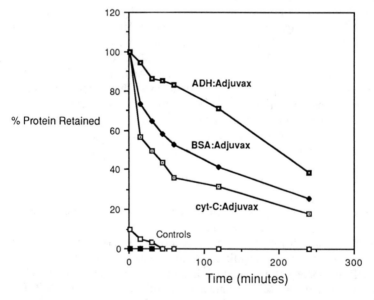

Figure 2. Effect of ligand molecular weight on the sustained release rate from Adjuvax.

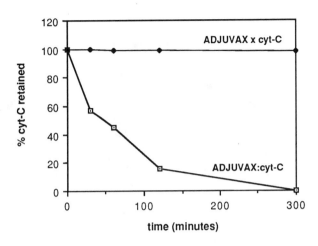

Figure 3. Effect of Adjuvax-ligand chemical crosslinking on ligand sustained release rate.

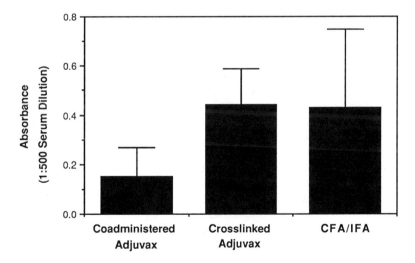

Figure 4. Comparison of Freund's Adjuvant to Adjuvax formulations in stimulating antibody response to P55 oligopeptide antigen. Relative antibody titers between adjuvant groups at day 27 were determined by measuring the absorbance at 450 nm of a 1:500 dilution of anti-P55 immune sera by ELISA.

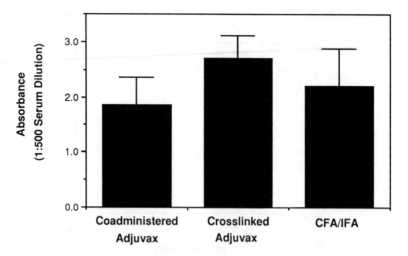

Figure 5. Comparison of Freund's Adjuvant to Adjuvax formulations in stimulating antibody response to BSA. Relative antibody titers between adjuvant groups at day 27 were determined by measuring the absorbance at 450 nm of a 1:500 dilution of anti-BSA immune sera by ELISA.

In addition, Adjuvax protects an antigen from degradation and rapid clearance from circulation through entrapment and sustained release from PGG glucan microcapsules. Diffusional release of entrapped proteins and peptides from the Adjuvax microcapsule was shown to be dependent upon molecular branching within the capsule matrix and protein/peptide molecular weight. It was also shown that covalent crosslinking of peptides or proteins to the Adjuvax decreases the release rate to the extent that release becomes dependent upon in vivo biodegradation of the crosslinking bonds rather than diffusion.

In vivo adjuvant studies with peptide and protein loaded Adjuvax, crosslinked Adjuvax-antigen conjugates, or the standard CFA adjuvant demonstrate that the three formulations stimulate comparable antibody responses in mice. The trends observed showed that Adjuvax-antigen loaded preparations yielded weaker antibody responses than CFA, while Adjuvax-antigen crosslinked preparations yielded stronger antibody responses than CFA, although there were no statistically significant differences between the CFA and Adjuvax groups. These results support the importance of the targeted antigen delivery features of Adjuvax which require the prolonged association between the antigen and the Adjuvax microcapsule to obtain the greatest stimulation of antibody titers.

It was also shown that, unlike CFA and many other adjuvants, Adjuvax itself is nonantigenic (data not shown) allowing repeated use of Adjuvax-based drug or vaccine formulations with minimal side-effects. In addition, unlike many other oil or detergent based adjuvants, the saline based Adjuvax does not cause local pathologies, such as granulomas at the injection site, and the effectiveness of Adjuvax to stimulate antibody titers also has been established in a wide variety of animal species ranging from mice to horses (data not shown). These combined efficacy and safety features of Adjuvax support its further evaluation and development for use as an adjuvant or drug delivery vehicle for human and veterinary applications.

Literature Cited

1. Jamas S, Easson DD, Ostroff GR, Onderdonk AB. 1990. ACS Preprint.
2. Czop, J. K.; Austen, K. F. *J. Immunol.* **1985**, *134*, 2588.
3. Czop, J.; Fearon D. T.; Austen, K.F. *J. Immunol.* **1978**, *120*, 1132.
4. Czop, J.K.; Puglisi, A.V.; Miorandi, D.M.; Austen, K.F. *J. Immunol.* **1988**, *141*, 3170.
5. Williams, J.D.; Czop, J.K.; Austen, K.F. *J. Immunol.* **1984**, *132*, 3034.
6. Czop, J.K.; Austen, K.F. *Proc. Natl. Acad. Sci. USA* **1985**, *82*, 2751.
7. Janusz, M.J.; Austen, K.F.; Czop, J.K. *J. Immunol.* **1987**, *138*, 3897.
8. Benach, J.L.; Habicht, G.S.; Holbrook, T.W.; Cook, J.A. *Infect. Immun.* **1982**, *35*, 947.
9. Cook, J.A.; Holbrook, T.W.; Dougherty, W.J. *Infect. Immun.* **1982**, *37*, 1261.
10. Holbrook, T.W.; Cook, J.A.; Parker, B.W. *Infec. Immun.* **1981**, *32*, 542.
11. Ostroff, G.R.; Easson, D.D.; Jamas, S. 198th ACS National Meeting. September 1989.
12. Symposium: The Beta-Glucan Receptor and Response to PGG, International Congress for Infectious Diseases, Montreal Canada, July 1990.

RECEIVED March 19, 1991

Chapter 7

Fluorescent Labeling of Heparins and Related Polysaccharides

Old Problems and New Solutions

Michael Sobel[1], Raphael M. Ottenbrite[2], and Yasuo Suda[3]

[1]Department of Surgery, Medical College of Virginia, Virginia
Commonwealth University, Richmond, VA 23298–0108
[2]Department of Chemistry, Virginia Commonwealth University,
Richmond, VA 23298–0108
[3]Department of Chemistry, Faculty of Science, Osaka University, Toyonaka,
Osaka 560, Japan

Using a new synthetic fluorescein-containing reagent, we have successfully labelled heparins via their hydroxyl groups, under mild pH, ionic strength, and aqueous conditions. The efficiency of labelling is high, producing a degree of substitution of over one molecule of fluorescein per compound. The degree of labelling can be precisely controlled by the concentrations of the reactants, and the fluorophore can be secondarily labelled with [125]Iodine. The fluorescent labelled heparin retained its affinity for antithrombin, as measured by fluorescence polarization, and by anticoagulant assays. This new reagent may be very useful for the selective and controlled labelling of a wide range of bioactive polysaccharides and proteins where attachment via amino groups is not possible or desirable.

Nondestructive techniques for the labelling of glycosaminoglycans are essential for their study in vitro and in vivo. The labelling of heparins is especially problematic. First, they lack intrinsic chromophores, markers or other sites for simple labelling, and secondly the relationship between structure and biologic function is especially complex, making even seemingly minor alterations of structure detrimental to function. Historically, heparins have been labelled with ^{3}H isotopes by direct hydrogen exchange, by acetylation with [^{14}C]acetic anhydride (1), or by partial degradation and reduction with [^{3}H]borohydride (2). As the field of heparin research has expanded, so has the need for more versatile labelling methods.

The nondestructive introduction of a fluorescent label would provide the molecule with a nonradioactive fluorophore, yet would preserve the option for direct radiolabelling of the fluorescent moiety with [125]Iodine. This approach was pioneered by Nagasawa et al. (3) who reacted native or N-desulfated heparins with a fluorescein isothiocyanate (FITC). The resulting degree of labelling was low

0097–6156/91/0469–0060$06.00/0

(~0.1fluorescein/heparin) and was highly dependent upon the availability of unsubstituted hexosamine sugars. More extensive labelling could be achieved by *N*-desulfation, but at the expense of biologic anticoagulant activity. Jordan and Rosenberg (*4*) directly reacted heparin with fluorescamine to yield a fluorescent heparin product. They demonstrated that the labelling procedure did not alter the product's binding behavior with antithrombin. Although useful for analytic studies, this method had limited applicability because (1) only micro- to nanomolar quantities of heparin were labelled, (2) the fluorescent labelling reagent was not water soluble, and (3) a relatively low intensity of labelling was obtained. Subsequently, Ogamo et al. (*5*) utilized an intermediate carbodiimide condensing reagent to label the carboxyl groups of heparin through the amino group of 5-aminofluorescein. This technique offered higher intensities of labelling (over 1 fluorescein/heparin), but required harsher chemical conditions (pH 4.75) and more complex procedures, including an intermediate reagent. This method still relied upon an important functional group of heparin (-COOH) that may be essential for biologic activity (*6*).

Glabe et al. (*7*) proposed activation of the heparin polysaccharide with cyanogen bromide, and subsequent covalent attachment of fluorescamine. The cyanogen bromide chemistry preferentially attacks the -OH and $-NH_2$ groups of the polysaccharide. They achieved degrees of labelling up to 0.4 molecules of fluorescein per heparin molecule. The disadvantages of this method are the possible destructive effects of cyanogen bromide on the polysaccharide, the selective labelling of the unsubstituted amino groups, and the difficulty in precisely controlling the reaction. Reliance upon the rare unprotonated amino groups of heparin for label attachment has been criticized by Jones et al. (*8*), who observed that such labelling methods preferentially labelled a subpopulation of heparins with relatively poor capacity for binding antithrombin.

It is our opinion that hydroxyl groups may be preferable for the attachment of the label, because they exist predominantly outside of the tertiary structure of the biopolymer in aqueous solution, and the labelled chromophore bound to the hydroxyl group may least influence the tertiary structure and biological activity of the biopolymer. A reagent which was both fluorescent and reacted directly with the polysaccharide in a single step reaction would be simpler and more convenient. Finally, the reagent should be water soluble. In a recent report (*9*) we described the chemical synthesis and preliminary application of a new fluorescein reagent which fulfills many of these requirements: 2-(2,4-diazobicyclo-2,2,2-octyl)-4-(5-aminofluoresceinyl)-6-morpholinyl 1,3,5-triazine (called F-D and illustrated in Figure 1).

Experimental

Preparation of Reagent and Labelling Procedures. The structure of F-D [2-(2,4-diazobicyclo-2,2,2-octyl)-4-(5-aminofluoresceinyl)-6-morpholinyl 1,3,5-triazine] has been confirmed by its FAB-MS, IR, and ^{1}H-NMR spectra (*9*). Briefly, F-D was synthesized by the treatment of fluorescamine isomer I with cyanuric chloride, then reaction with morpholine and DABCO (1,4-diazobicyclo-2,2,2-octane), as illustrated

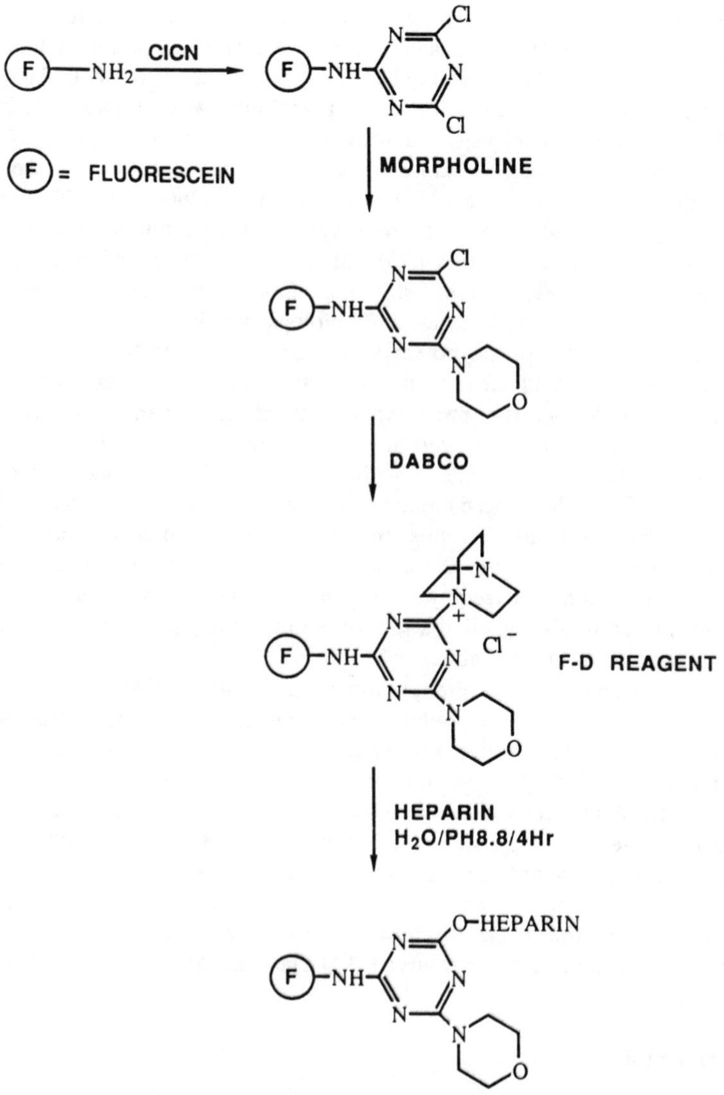

Figure 1. Synthesis of F-D and predicted reaction with heparin.

in Figure 1. The resulting water soluble reagent contains a fluorescein moiety as well as a reactive leaving group which can be directly displaced with an -OH group of the polysaccharide to form a covalent fluorescent polysaccharide with a one step labelling procedure (Figure 1).

In some situations, the direct attachment of a large group (such as fluorescamine) to a biologically active substrate can reduce activity. This is due to steric hindrance which can cause a change in conformation or physically block an active site. This condition can be obviated in many cases by attaching the bulky moiety to a spacer arm composed of two or more methylene groups. To this end, a variant of F-D was also synthesized in which a spacer arm of beta-alanine was inserted between the fluorescent and chemically reactive moieties of the reagent.

A typical labelling reaction is performed as follows: purified heparin (25mg) is dissolved in 1 mL of 0.2 M sodium bicarbonate buffer (pH 8.4) containing F-D (0.1 to 1 mg/mL). The reaction solution is shaken or gently stirred at 25°C for 5 hours or at 4°C overnight. D-glucose (10 mg) is added to react with excess reagent, and then the labelled heparin is separated from the reaction products by chromatography in water on a 1 x 18 cm Biorad P-2 column. The labelled heparin is pooled from the void volume and lyophilized. To determine the degree of labelling of the product, the heparin content is determined by uronic acid assay (*10*) and the content of fluorescein determined using a standard calibration curve developed from known concentrations of fluorescein isomer I in 0.2 M sodium bicarbonate buffer corrected for quenching by heparin. The percent labelling is generally expressed as a ratio of moles of fluorescein to moles of heparin or uronic acid in the labelled compound.

Results

Influences of Time, pH, and Molecular Weight on Labelling Efficiency. The degree of fluorescent substitution of the labelled heparin was expressed as a molar ratio of fluorescein to uronic acid. Maximal labelling of heparin was achieved in 5 hours at room temperature. A porcine mucosal heparin (MW 12,000) was labelled over a range of pH values. Figure 2 shows that labelling efficiency increased with rising pH - primarily the result of the increased solubility of F-D at higher pH. Labelling efficiency was independent of the polysaccharide chain length of the heparin. Shown in Figure 3 are the results of labelling heparins of MW 3,000, 4,000-6,000 and 12,000-15,000 (these were prepared by nitrous acid depolymerization of porcine mucosal heparin). A single linear relationship exists between the intensity of labelling of the final product and the initial ratio of fluorescein reagent to monosaccharide unit. Practically speaking, knowledge of this quantitative relationship permits precise control over the intensity of labelling of polysaccharides or heparins of any molecular weight range.

Investigation of the Site of Fluorescent Labelling. It was our goal with this new reagent to avoid the preferential labelling via -NH$_2$ groups. Therefore we pursued a series of experiments designed to (a) determine the dependency of F-D labelling on free -NH$_2$ groups, and (b) compare F-D with a method known to prefer -NH$_2$ groups

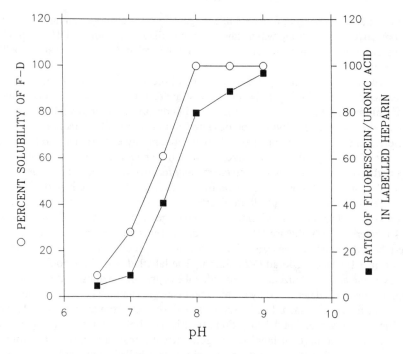

Figure 2. The pH dependency of heparin labelling. The degree of labelling intensity of the final product was compared with the concomitantly measured solubility of the F-D labelling reagent.

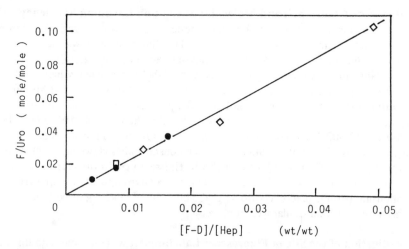

Figure 3. Effects of molecular weight on labelling efficiency. Three heparins were independently labelled at increasing degrees of substitution. Molecular weight ranges of 3,000 (open square), 4,000-6,000 (solid circle), and 12,000-15,000 (open diamond).

(fluoresceinisothiocyanate, FITC). The amino groups on heparin were N-blocked with N-hydroxysuccinimidylacetate (NSA). This reagent was prepared by the reaction of N-hydroxysuccinimide and acetic acid in the presence of N,N'-dicyclohexylcarbodiimide, and its structure confirmed by its IR and NMR spectral data. NSA was reacted with bovine lung heparin (25 mg/mL) at several different ratios of concentration (0 - 0.78 mg NSA/mg heparin). The reaction mixture was dialyzed and lyophilized to provide N-blocked heparin. IR spectra confirmed that the amino content of heparin decreased correspondingly after reaction with increasing amounts of NSA.

These variably N-blocked heparins were then independently fluorescent labelled with F-D or FITC to a high, medium or low extent by reaction with different concentrations of labelling reagent. Figures 4a and 4b show that the labelling of the N-blocked heparin with FITC was clearly dependent on the availability of free amino groups - the intensity of labelling dropped dramatically with increasing N-blockade. By contrast, the labelling efficiency of F-D was about one order of magnitude higher than FITC at all concentrations, and showed little dependence on pre-treatment with NSA.

Applications of Fluorescent Labelled Heparin in Experimental Biology. To assess the binding of labelled heparins to human antithrombin, the fluorescence polarization of mixtures of the fluorescent heparin and antithrombin were measured and calculated according to the following equation for anisotropy:

$$\text{Anisotropy} = (I_{vv}-(G)I_{vh})/(I_{vv}-2(G)I_{vh})$$

(1)

where I_{vv} and I_{vh} represent the emission intensities at 520nm (ex. 490 nm) parallel and perpendicular to the plane of polarization, respectively, and G is a correction factor for the transmission efficiency of the emission monochrometer to light polarized parallel and perpendicular to the grating (adapted from Bentley et al. (*11*)). The results are shown in Figure 5. In contrast to the interaction with bovine serum albumin, the values of anisotropy of the labelled heparin rose as the ratio of anti-thrombin to heparin (moles/mole) increased, reaching a maximum at a ratio of 1/1. We also assayed heparins labelled with F-D to determine their anticoagulant activity, using a bioassay for anti-factor Xa activity (*12*). The data in Table I confirm that heparins labelled with the F-D reagent retain their biologic anticoagulant activity over a range of labelling intensities. Heparins labelled with the F-D reagent prepared with a beta-alanine spacer arm also showed normal anticoagulant activity (data not shown).

One of the first practical applications for these fluorescent labelled heparins was to examine the heparin binding behavior of different proteins and peptides under study in our laboratories. To this end we used a modification of the dot-blot assay described by Hirose and colleagues (*13*). F-D labelled heparin (~1 fluorescein/heparin) was radiolabelled with [125]Iodine using iodobeads, to a specific activity of approximately 0.5×10^6 cpm/µg. Solutions of proteins with known heparin-binding capacities were dotted on nitrocellulose paper. A series of replicates

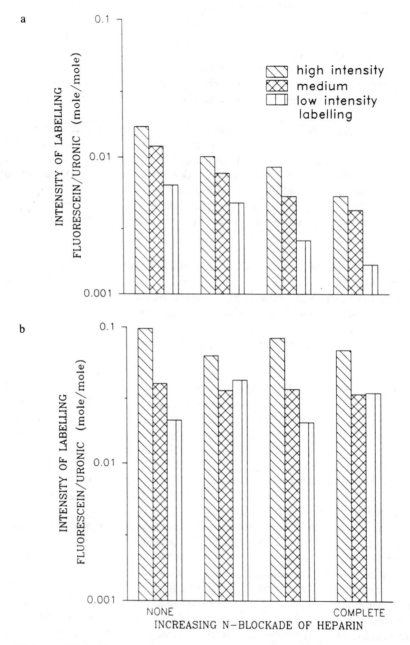

Figure 4. The efficiency of labelling of N-blocked or nonblocked heparins with FITC versus F-D. See text for details of experiment: a, FITC; b, F-D.

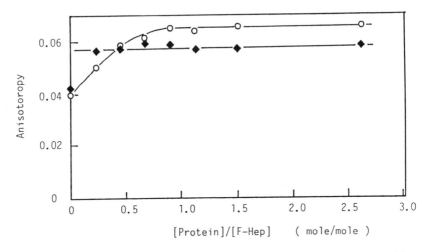

Figure 5. Fluorescence anisotropy of F-D labelled heparin-antithrombin interaction. F-D-heparin (0.02 fluoresceins per uronic acid) at 0.1 mg/ml was incubated with different concentrations of antithrombin (open circles) or bovine serum albumin (solid diamonds) in 20 mM sodium phosphate buffer, pH 7.4.

Table I. Intensity of Labelling versus Heparin Anticoagulant Activity

Percent Labelled (moles F/mole heparin)	Anticoagulant Activity (anti-factor Xa units)
126%	150 units/mg
41%	198 units/mg
no label	187 units/mg

were independently incubated with heparin solutions of increasing concentration, washed, dried and autoradiogrammed. Figure 6 shows that the ^{125}I-F-D-heparin quantitatively bound to antithrombin (AT-III), von Willebrand factor (VWF), and a synthetic heparin binding peptide (P-III), but not to ovalbumin (OVALB), the negative control. Binding was saturable and competitive as unlabelled heparin displaced the binding of ^{125}I- F-D-heparin.

Discussion

These data demonstrate a new, practical method for introducing a fluorescent label into polysaccharides via their hydroxyl groups. The reaction can be performed under aqueous conditions of mild pH and ionic strength. The intensity of labelling can be finely controlled, and even high degrees of substitution do not appear to significantly alter the biologic activities of the labelled compounds.

Maximum labelling of heparin with F-D was achieved at 5 hours at 25°C, pH 8.4. In the case of heparin, the efficiency of labelling was not dependent on molecular weight, but solely a function of the ratio of the concentrations of labelling reagent to monosaccharide subunit in the reaction mixture. Similar results were encountered in the labelling of dextrans of different molecular weight (9).

As expected, the labelling of heparin with FITC appeared to be dependent upon the content of free amino groups, as judged by the reduced efficiency of labelling with N-blocked heparins. The novel fluorescein derivative, F-D, demonstrated an almost tenfold higher labelling efficiency, which was independent of the availability of free amino groups. These observations suggest that the F-D reacts primarily with available hydroxyl groups in the heparin rather than amino groups. With rising pH, the efficiency of labelling increased in parallel with the solubility of the labelling reagent (Figure 2). The pKa value of the amino group is approximately 9, and therefore heparin should experience a net increase in the number of activated unprotonated amino groups at the higher pH region. At such a high pH it is likely that F-D would react with both hydroxyl and amino groups, and therefore we have chosen to perform labelling at a pH below the pKa of the amino groups.

The fluorescent labelling of heparin with F-D by this technique did not observably alter the biologic activity of the heparin as regards to its binding to antithrombin and catalysis of antithrombin's neutralization of activated coagulation factors. F-D labelled heparins also bound to other known heparin-binding proteins in a saturable and reversible manner, as demonstrated by the dot-blot assay technique (Figure 6).

In all cases, the labelling reactions were performed easily and conveniently, and the final product separated to high purity with the simplest chromatographic methods. The isotope ^{125}I was easily introduced to the F-D moiety at a second step. This new fluorescent labelling reagent may have applicability to a potentially large group of polysaccharides and proteins which lack free amino groups or in which derivatization via such groups may be deleterious to biologic activity. Proteins without an available tyrosine may be labelled through the -OH groups of serine or threonine. Finally,

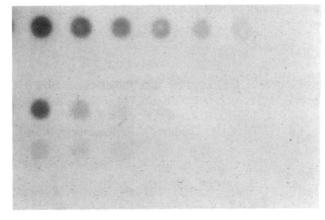

AT-III
OVALB
P-III
VWF

AT-III
OVALB
P-III
VWF

Figure 6. ^{125}I-F-D Heparin dot blot assay. See text for methods and abbreviations. Upper panel: incubation with ^{125}I-F-D labelled porcine mucosal heparin alone, Lower panel: the same conditions, but a 100-fold excess of unlabelled heparin has been added to the labelled heparin.

these chemical approaches may be useful for the development of new methods for the covalent attachment of compounds to solid matrices and substrates.

Acknowledgments

The authors are grateful to Ms. Dalila Marques for her expert research assistance. This work was supported by NIH grant HL39903 and the Veterans Administration Merit Review Agency.

Literature Cited

1. Jordan, R.E.; Beeler, D.; Rosenberg, R.D. *J. Biol. Chem.* **1979**, *254*, 2902.
2. Tollefsen, D.M.; Peacock, M.E.; Monafo, W.J. *J. Biol. Chem.* **1986**, 858.
3. Nagasawa, K.; Uchiyama, H. *Biochim. Biophys. Acta* **1978**, *544*, 430.
4. Jordan, R.E.; Oosta, G.M.; Gardner, W.T.; Rosenberg, R.D. *J. Biol. Chem.* **1980**, *255*, 10073.
5. Ogamo, A.; Matsuzaki, K.; Uchiyama, H.; Nagasawa, K. *Carbohydr. Res.* **1982**, *105*, 69.
6. Danishefsky, I.; Ahrens, M.; Klein, S. *Biochim. Biophys. Acta* **1977**, *498*, 216.
7. Glabe, C.G.; Harty, P.K.; Rosen, S.D. *Anal. Biochem.* **1983**, *130*, 287.
8. Jones, G.R.; Hashim, R.bt.; Power, D.M. *Biochim. Biophys. Acta* **1987**, *925*, 57.
9. Suda, Y.; Sobel, M.; Sumi, M.; Ottenbrite, R.M. J. *Bioact. Compatible Polymers* **1990**, *5*, 412.
10. Bitter, T.; Muir, H.M. *Anal. Biochem.* **1962**, *4*, 330.
11. Bentley, K.L.; Klebe, R.J.; Hurst, R.E.; Horowitz, P.M. *J. Biol. Chem.* **1985**, *260*, 7250.
12. Teien, A.N.; Lie, M. *Thromb. Res.* **1977**, *10*, 399.
13. Hirose, N.; Krivanek, M.; Jackson, R.L.; Cardin, A.D. *Anal. Biochem.* **1986**, *156*, 320.

RECEIVED April 1, 1991

Chapter 8

Design of Polysaccharide-5-Fluorouracil Conjugates Exhibiting Antitumor Activities

T. Ouchi, T. Banba, T. Z. Huang, and Y. Ohya

Department of Applied Chemistry, Faculty of Engineering, Kansai University, Suita, Osaka 564, Japan

In order to provide a macromolecular prodrug of 5-fluorouracil(5FU) with reduced side-effects, having affinity for tumor cells and exhibiting a high antitumor activity, the design of polysaccharide-5FU conjugates was investigated. Chitin-5FU, chitosan-5FU, α-1,4-polygalactosamine-5FU, partially N-acetylated α-1,4-polygalactosamine-5FU, hyaluronic acid-5FU and dextran-5FU conjugates exhibited significant survival effects against p388 lymphocytic leukemia in mice by intraperitoneal(ip) transplantation/intraperitoneal(ip) injection. Chitosan-5FU, chitosamino-oligosaccharide-5FU, and galactosamino-oligosaccharide-5FU conjugates showed higher growth-inhibitory effects against MH134Y hepatoma and Meth-A fibrosarcoma in mice than 5FU, chitin, oligosaccharides and their blends by subcutaneous(sc) implantation/intravenous(iv) injection. The obtained conjugates did not display an acute toxicity in the high dose ranges.

5-Fluorouracil(5FU) has a remarkable antitumor activity (*1-3*), which is accompanied, however, by undesirable side-effects (*4,5*). Polysaccharides such as chitin, chitosan, α-1,4-polygalactosamine, N-acetyl-α-1,4-polygalactosamine, hyaluronic acid and dextran are noteworthy as low or nontoxic, nonimmunogenetic, compatible and biodegradable polymers. Partially N-acetylated chitosan and α-1,4-polygalactosamine have been further reported to be selectively collected into tumor cells (*6*) and to inhibit growth of tumor cells (*7*), respectively. Especially, water-soluble hexa-N-acetyl chito-hexaose and α-1,4-polygalactosamine were found to act as growth-inhibitors of tumors through stimulation of the host immuno-system (*8*).

So, in order to provide a macromolecular prodrug of 5FU with reduced side-effects, having affinity for tumor cells and exhibiting a high antitumor activity, the present paper is concerned with the design of hybrid type conjugates of chitin-

0097–6156/91/0469–0071$06.00/0

5FU($\underline{1}$), chitosan-5FU($\underline{2}$), α-1,4-polygalactosamine-5FU($\underline{3}$), partially N-acetylated α-1,4-polygalactosamine-5FU($\underline{4}$), hyaluronic acid-5FU($\underline{5}$), dextran-5FU($\underline{6}$), and 6-0-carboxymethyl chitin(CM-chitin)-5FU($\underline{7}$).

Experimental

Syntheses of Polysaccharide-5FU Conjugates. The conjugates of 5FUs attached to chitosan, α-1,4-polygalactosamine, partially N-acetylated α-1,4-polygalactosamine and dextran at the 2-positions through hexamethylene spacer groups via carbamoyl bonds were synthesized by coupling reactions of chitosan, α-1,4-polygalactosamine, partially N-acetylated α-1,4-polygalactosamine and dextran with 6-(5-fluorouracil-1-yl)-hexamethylene isocyanate($\underline{8}$) (9), respectively. Chitosaminohexaose(COS6)-5FU($\underline{9}$), chitosaminotriose(COS3)-5FU($\underline{10}$), galactosaminopentaose (GOS5)-5FU($\underline{11}$) and galactosaminotriose(GOS3)-5FU($\underline{12}$) conjugates were prepared by the same methods described above. The desired conjugates of 5FU attached to chitin and hyaluronic acid at 6-positions through hexamethylene and pentamethylene spacer groups via carbamoyl bonds were synthesized through coupling reactions of chitin and hyaluronic acid with the isocyanate ($\underline{8}$) and 1-[(5-aminopentyl)carbamoyl]- 5-fluorouracil ($\underline{13}$) (10), respectively. CM-chitin-5FU conjugate ($\underline{7}$) was prepared by coupling reaction of CM-chitin with 4-[(amino-n-pentyl)-ester]-methylene-5FU($\underline{14}$) (11). These fixation steps of 5FU onto polysaccharides are shown in Schemes I-IV.

Release Behavior of 5FU from Polysaccharide-5FU Conjugates. In order to evaluate the release behavior of 5FU from the polysaccharide-5FU conjugates, the hydrolyses of these conjugates were studied in vitro at 37°C in physiological saline. The carbamoyl and ester bonds of the obtained polysaccharide-5FU conjugates were cleaved to give only free 5FU itself but not to afford any 5FU derivative. After the release of 5FU, the glycoside bonds of backbone polysaccharides might be degraded slowly for a very long time.

Survival Effect of Polysaccharide-5FU Conjugates against p388 Leukemia in Mice ip/ip. The survival effect of the 5FU conjugates was evaluated against p388 lymphocytic leukemia in female CDF_1 mice (30 untreated mice/group and 6 treated mice/group) by ip transplantation/ip injection according to the protocol for preliminary screening of the antitumor activity performed at the Japanese Foundation for Cancer Research. A total of 1×10^6 leukemia cells were injected ip on day 0. The water-insoluble chitin-5FU($\underline{1}$), chitosan-5FU($\underline{2}$), α-1,4-polygalactosamine- 5FU($\underline{3}$), partially N-acetylated α-1,4-polygalactosamine- 5FU($\underline{4}$) and dextran-5FU($\underline{6}$) conjugates were sonicated in 0.05% Sorbate 80 and sterile normal saline, and then the obtained suspensions were administered ip. Hyaluronic acid-5FU($\underline{5}$) and CM-chitin-5FU($\underline{7}$) conjugates were dissolved in sterile normal saline, and then the obtained aqueous solutions were administered ip. The mice received doses of 200-800mg/kg on days 1 and 5. The ratio of prolongation of life, T/C(%), which means the ratio of median survival of treated mice (T) to that of control mice (C) was evaluated as the survival effect.

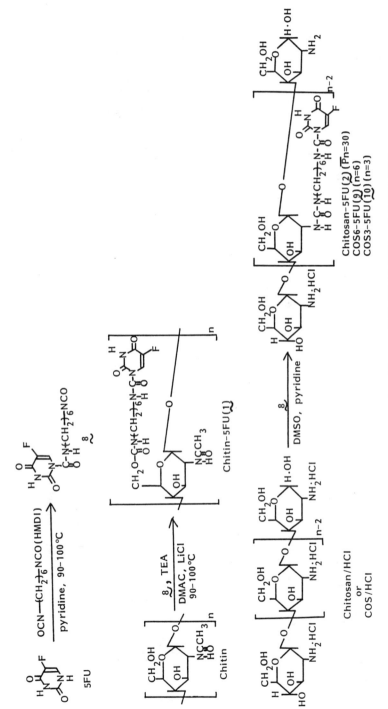

Scheme I. Preparation route for chitosan-5FU, COS6-5FU and COS3-5FU conjugates.

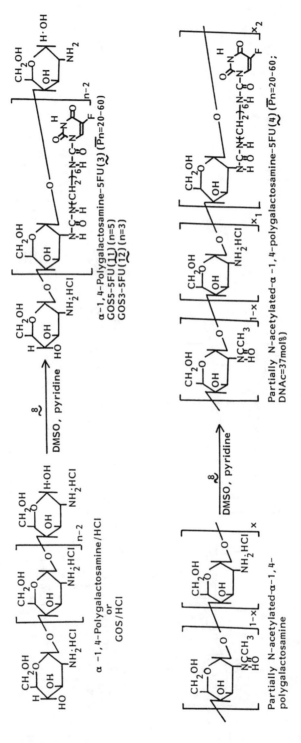

Scheme II. Preparation routes for α-1,4-polygalactosamine-5FU, GOS5-5FU, GOS3-5FU conjugates and partially N-acetylated-α-1,4-polygalactosamine-5FU conjugates.

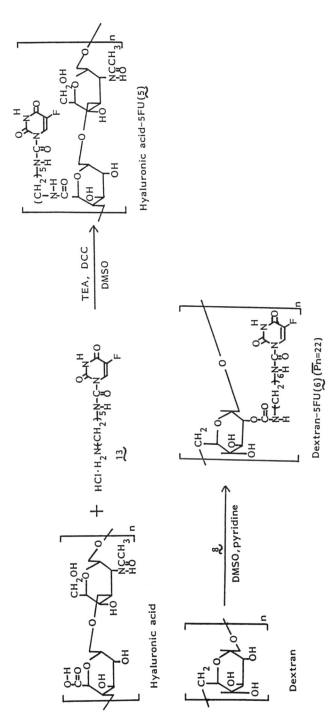

Scheme III. Preparation routes for hyaluronic acid-5FU and dextran-5FU conjugates.

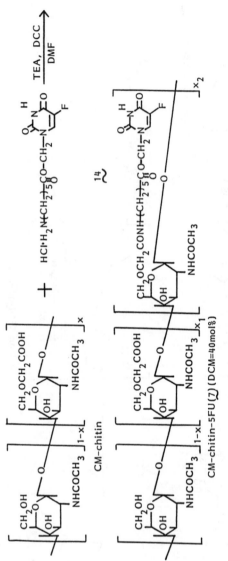

Scheme IV. Preparation routes for CM-chitin-5FU conjugate.

Growth-inhibitory Effect of Polysaccharide-5FU Conjugate against Solid Tumor Cells in Mice sc/iv. The growth-inhibitory effect was tested against MH134Y hepatoma or Meth-A fibrosarcoma in male mice (6 or 10 untreated mice/group and 6 or 10 treated mice / group) by sc implantation/iv injection. The water-insoluble chitosan-5FU(2), COS6- 5FU(9), COS3-5FU(10), α-1,4-polygalactosamine-5FU(3), GOS5-5FU(11) and GOS3-5FU(12) conjugates dissolved in a small amount of dimethyl sulfoxide (DMSO) were suspended in sterile normal saline or buffer solution and then the obtained suspensions were administered iv. Either 2×10^5 or 5×10^5 MH134Y hepatoma cells or Meth-A fibrosarcoma cells were injected sc on day 0. The mice received doses of 1 or 10 mg/kg of oligosaccharide-5FU conjugate on day 14 (one time) or days 7, 12, and 17 (three times). The mice were sacrificed after 30 days and the solid tumor weights were measured. The growth-inhibitory effects of these polysaccharide-5FU and oligosaccharide-5FU conjugates against the solid tumor cells were evaluated by the ratio of the difference between the tumor weight of the test animals and the control animals(W_C-W_T) compared to that of the control animals (W_C) after 30 days.

Results and Discussion

Survival Effects. The results of the survival effects for chitin-5FU(1), chitosan-5FU(2), α-1,4-polygalactosamine-5FU(3), partially *N*-acetylated α-1,4-polygalactosamine(4), hyaluronic acid- 5FU(5), dextran-5FU(6), and CM-chitin-5FU(7) conjugates against p388 leukemia in female CDF mice ip/ip are summarized in Figures 1-5. Chitin-5FU, chitosan-5FU, α-1,4-polygalactosamine-5FU, partially *N*-acetylated α-1,4- polygalactosamine-5FU, hyaluronic acid-5FU, CM-chitin-5FU conjugates were found to exhibit higher level survival effects than free 5FU against p388 leukemia mice. The survival effect tended to increase with increasing the degree of substitution of 5FU based on the number of sugar groups (D5FU). Moreover, the passive targeting of polysaccharide-5F conjugate to specific organs was suggested to be achieved by the balance of electric charge and hydrophilicity / hydrophobicity.

Growth-Inhibitory Effects. The growth-inhibitory effect of chitosan-5FU conjugate (2) against Meth-A fibrosarcoma in SPF-C3H/He male mice was obtained as shown in Figure 6(A). Figure 6(B) shows the growth-inhibitory effect of COS6-5FU conjugate(9) against MH134Y hepatoma in SPF-C3H/He mice sc/iv. Figure 6(C) shows the growth-inhibitory effect of COS3-5FU conjugate(10) against Meth-A fibrosarcoma in Balb/c male mice sc/iv. From the results shown in Figure 6, these chitosan-5FU and COS-5FU conjugates were found to exhibit higher growth-inhibitory effects against these solid tumors than 5FU, chitosan, COS and their blends. On the other hand, the growth-inhibitory effects of GOS5-5FU(11) and GOS3-5FU(12) conjugates against Meth-A fibrosarcoma in Balb/c male mice sc/iv are shown in Figure 7. GOS3-5FU conjugate(12) was found to exhibit the highest level of growth-inhibitory effect.

Moreover, chitin-5FU, chitosan-5FU, COS-5FU, α-1,4- polygalactosamine-5FU, partially *N*-acetylated α-1,4- polygalactosamine-5FU, GOS-5FU, hyaluronic

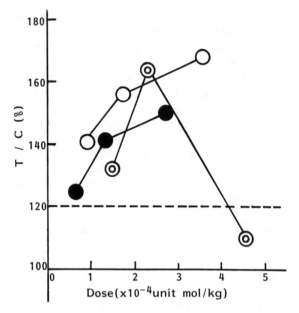

Figure 1. Survival effects for chitin-5FU conjugates(1) against p388 leukemia in mice ip/ip. ● : D5FU=8.3mol%, ◯ : D5FU=11mol%, ◉: 5FU.

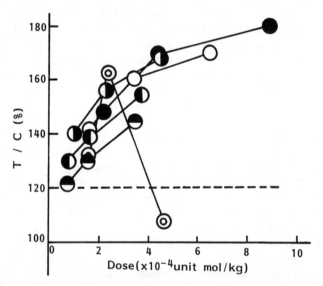

Figure 2. Survival effect for chitosan (Pn=30)-5FU conjugate(2) against p388 leukemia in mice ip/ip. ◑: D5FU=7.0mol%, ◐: D5FU=8.3mol%, ◑: D5FU=11mol%, ◯ :D5FU=18mol%, ● :DFU=27mol%, ◉ : 5FU.

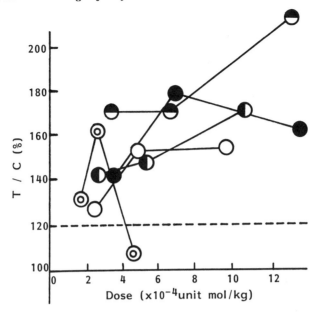

Figure 3. Survival effect for α-1,4-polygalactosamine-5FU(3) and partially *N*-acetylated α-1,4-polygalactosamine-5FU(4) conjugates against p388 leukemia in mice ip/ip. ◯ : 3(Pn=20-60; D5FU=28mol%), ◐ : 3(Pn=20-60; D5FU=34mol%), ◓ : 3(Pn=20-60; D5FU=47mol%), ● : 4(Pn=20-60; DNAc[a]=37mol% D5FU=52mol%), ⊙ :5FU. [a]Degree of substitution of *N*-acetyl groups based on the number of sugar groups.

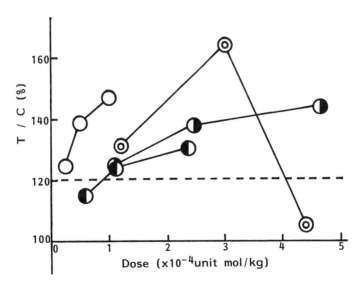

Figure 4. Survival effects for hyaluronic acid-5FU(5) and dextran-5FU(6) conjugates against p388 leukemia in mice ip/ip. ◯ : 5(D5FU=5mol%), ◑ : 6(D5FU=6mol%), ◐ : 6(D5FU=11mol%), ⊙ : 5FU.

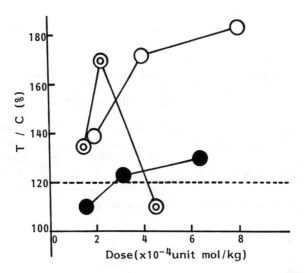

Figure 5. Survival effects for CM-chitin-5FU conjugate(7) and the hydrochloride salt of 4-[(amino-*n*-pentyl)ester]methylene-5FU(14) against p388 leukemia in mice ip/ip. ○ : 7(DCM^a=40mol%; D5FU=35mol%), ● :14, (○) : 5FU. aDegree of substitution of carboxymethyl groups based on the number of sugar groups.

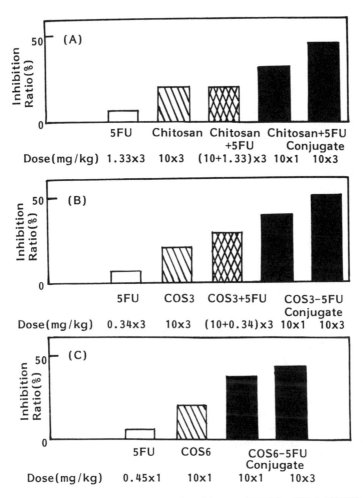

Figure 6. Growth-inhibitory effects for chitosan-5FU(2), COS6-5FU(9) and COS3-5FU(10) conjugates against MH134Y hepatoma or Meth-A fibrosarcoma in mice sc/iv. (A): 2(Pn=30); D5FU=27mol%), (B): 9(D5FU=34mol%), (C): 10(D5FU-44mol%).

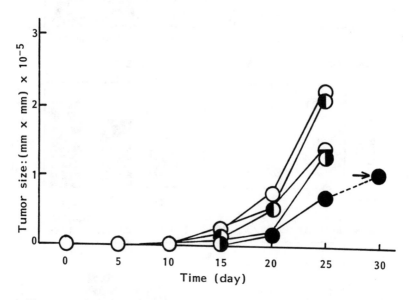

Figure 7. Growth-inhibitory effects for GOS5-5FU(*11*) and GOS3-FU(*12*) conjugates against Meth-A fibrosarcoma in mice sc/iv. ◯: control, ◖: *11*(D5FU=58mols%) 1mg/kg × 3 iv, ◕: *11*(D5FU=58mol%) 10mg/kg × 3 iv, ●: *12*(D5FU=38mol%) 1mg/kg × 3 iv, ◑: *12*(D5FU=38mol%) 10mg/kg × 3 iv. → : Growth-inhibitory effect=64%.

acid-5FU and dextran-5FU conjugates did not cause a rapid decrease in body weights of treated mice in the high dose ranges. As a result they did not display an acute toxicity. Therefore, the macromolecule-5FU conjugate technique can be concluded to decrease significantly the side-effects of 5FU.

The results obtained suggest that such polysaccharide-5FU conjugates can be utilized as tumor therapeutical agents.

Acknowledgments

The authors wish to express their sincere appreciation to Dr. Tazuko Tashiro of Cancer Chemotherapy Center of the Japanese Foundation for Cancer Research for the screening test of the survival effect against p388 lymphocytic leukemia in mice ip/ip.

They also express thanks to Prof. S. Suzuki, Prof. M. Suzuki, Prof. T. Matsumoto of Tohoku College of Pharmacy for the measurement of growth-inhibitory effects against MH134Y hepatoma and Meth-A fibrosarcoma in mice sc/iv.

Literature Cited

1. Waxman, S.; Scanlon, K. J. In *Clinical Interpretation & Practice of Cancer Chemotherapy;* Greenspan, E. M., Ed.; Raven Press: New York, NY, 1982, pp.39.
2. Heidelberger, C. *Cancer Medicine, 2nd edn.,* Lea & Febiger: Philadelphia, PA, 1982, pp.801.
3. Myers, C. E. *Pharmacol. Rev.* **1981**, *1*, 33.
4. Bosch, L.; Harbers, E.; Heidelberger, C. *Cancer Rev.* **1958**, *18*, 335.
5. Bounous, G.; Pageau, R.; Regoli, D. *Int. J. Chem. Pharmacol. Biopharm.,* **1978**, *16*, 519.
6. Sirica, A. E.; Woodman, R. J. *J. Natl. Cancer Inst.,* **1971**, *47*, 377.
7. Ishiya, Y. et al., *Proceeding of 59th Annual Meeting of the Japanese Biochemical Society,* 1986, pp.871.
8. Suzuki, K.; Tokoro, A.; Okawa, Y.; Suzuki, S.; Suzuki, M. *Microbiol Immunol.* **1986**, *30*, 777.
9. Ouchi, T.; Banba, T.; Fujimoto, M.; Hamamoto, S. *Makromol. Chem.* **1989**, *190*, 1817.
10. Ouchi, T.; Kobayashi, H.; Banba, T. *Brit. Polym. J.* **1990**, *23*, 221.
11. Ohya, Y.; Kobayashi, H.; Ouchi, T. *Reactive Polymers,* submitted.

RECEIVED March 19, 1991

Chapter 9

Methotrexate-γ-Hydrazide and Divinyl Ether—Maleic Anhydride Conjugate

Antitumor Activity in Mice

Hwei-Ru Tsou, Roslyn Wallace, Ronald V. Citarella, and Janis Upeslacis

Oncology and Immunology Research Section, Medical Research Division, Lederle Laboratories, American Cyanamid Company, Pearl River, NY 10965

A water-soluble polymeric drug containing 26% of methotrexate-γ-hydrazide (MTX-hy) was synthesized by covalently coupling of MTX-hy to divinyl ether-maleic anhydride (MVE). The advantage of the polymeric drug (MVE-hy-MTX) is that drug attachment was not through the amino groups of the pteridine ring or the alpha-carboxyl group of the glutamic acid moiety, but rather through the amino group of the gamma-hydrazide of the glutamic acid moiety. MVE-hy- MTX was more active than MTX-hy against P388 leukemia in mice. Treatment with MVE-hy-MTX against human myeloid HL60 leukemia in athymic mice showed 96% tumor growth inhibition with no deaths at day 28 post treatment, while at the same dosage, MTX-hy produced 41% tumor inhibition with 83% survival. MVE-hy-MTX was equally active and approximately 8 to 10-fold less potent compared to free methotrexate in these studies.

Macromolecules have been used as drug carriers in an attempt to prolong plasma level of drugs presumably as a result of slow release of drugs from macromolecules and to achieve favorable uptake by the tumor cells (1). Among macromolecular carriers, divinyl ether-maleic anhydride (MVE) has been investigated extensively (2, 3). MVE copolymer contains multiple anhydride rings, which allows easy functionalization with antitumor agents carrying nucleophilic groups such as $-NH_2$, -OH and -SH. Furthermore, a carboxyl group is generated from each anhydride ring functionalized with a drug molecule. Therefore, MVE copolymer is capable of covalently binding a large number of lipophilic antitumor agents, while maintaining water solubility.

MVE copolymer has been covalently linked with 5-fluorouridine (4, 5), daunomycin (6), adriamycin (6-8), β-D-arabinofuranosylcytosine (9) and methotrexate (*10*). Most of the MVE-linked drugs demonstrated higher therapeutic efficacies and lower toxicities during in vivo antitumor evaluations. In previous studies methotrexate (MTX) was linked to MVE coplymer through the 2- or -4-amino groups of the pteridine ring. This MVE-linked MTX showed only a slightly enhanced increase in life span (%ILS) against L1210 leukemia in mice, when compared with free MTX (*11*). In our work, we linked MTX to MVE copolymer through its γ-hydrazino group to leave the 2- and 4-amino groups of the pteridine ring untouched, since these amines are thought to be important for inhibition of dihydrofolate reductase activity (*12, 13*). The antitumor activities of MVE-MTX-γ-hydrazide (MTX-γ-hy) were examined against P388 implanted in normal mice and HL60 in athymic mice.

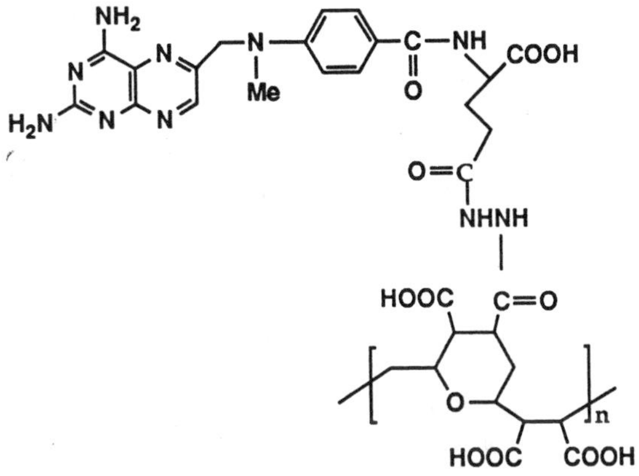

MVE-hy-MTX

Experimental

Material. Methotrexate-γ-hydrazide was prepared by a modified procedure described by Rosowsky (*14*). Divinyl ether-maleic anhydride copolymer was purchased from Hercules Incorporated (Wilmington, DE) and had an average molecular weight of 11,400. Centriprep-10 concentrators were purchased from Amicon (Danvers, MA).

Bio-Sil TSK- 250 gel filtration column (7.5mm x 30cm) was obtained from Bio-Rad (Richmond, CA), and used with 0.2 M phosphate (pH 4) mobile phase.

Preparation of MVE-methotrexate-γ-hydrazide. MVE copolymer (200 mg) was dissolved in 4 mL of N-methyl-2-pyrrolidinone. Triethylamine (0.38 mL) was added to the MVE solution, followed by methotrexate-γ-hydrazide (262 mg) in 4.5 ml of N-methyl-2-pyrrolidinone. The reaction solution was stirred at room temperature for three days.

The reaction solution was diluted with 36 mL of water. The unreacted free drug and other low molecular weight materials were removed by a Centriprep-10 concentrator. Purification was repeated until HPLC analysis (Bio-Sil TSK-250) of the product indicated the absence of free drug. The final purified product was lyophilized to afford 483 mg of a yellow solid. This solid product was readily soluble in water or aqueous buffer. The amount of drug covalently bound to MVE was estimated by the absorbance at 303 nm using a molar extinction coefficient of 17.03 x 10^3. The MVE-γ-hy-MTX contains 26% methotrexate-γ-hydrazide by weight.

Animal Studies. In the P388 murine leukemia tests, male BDFl mice weighing 18 to 21 g were injected intraperitoneally (ip) with 1 x 10^6 P388 tumor cells on day 0 of the test. Drugs were administered ip once daily for four consecutive days, starting one day after tumor inoculation. Six to twelve mice per group were used. The effect on survival was expressed as %ILS which was calculated as follows: ILS = [(T/C) - 1] x 100, where T/C is the median survival time (MST) of mice in the treated group (T) divided by the MST of the placebo treated control group (C). A value of %ILS equivalent to 35% or greater indicated positive drug activity.

In the human myeloid HL60 leukemia tests, athymic mice weighing 18 to 21 g were implanted subcutaneously with 2 x 10^7 HL60 tumor cells on day 0 of the test. Drugs were adminstered ip once daily for five consecutive days, starting one day after tumor implantation. Five to six mice per test group and ten mice per control group were used. Tumor weights were estimated from measurements of tumor diameters of treated (T) and the control (C) mice. A 58% reduction of tumor weight in treated mice compared to placebo treated controls indicated positive drug activity.

Results and Discussion

Methotrexate has two carboxyl groups on the glutamic acid moiety. Since α-amide derivatives of methotrexate showed a markedly lower inhibitory effect on dihydrofolate reductase than methotrexate, and γ-amide derivatives of methotrexate inhibited dihydrofolate reductase as effectively as methotrexate (*15,16*), we chose to convert the γ-carboxyl group of methotrexate to a hydrazide, which served as an anchor for attachment to the MVE copolymer. However, we observed that methotrexate-γ-hydrazide suffered 8 to 10-fold losses of antitumor activities against P388 leukemia and HL60 leukemia as compared to methotrexate. To understand the effect of drugs conjugated to polymers, we compared the antitumor activities between MVE-γ-hy-MTX and MTX-γ-hy.

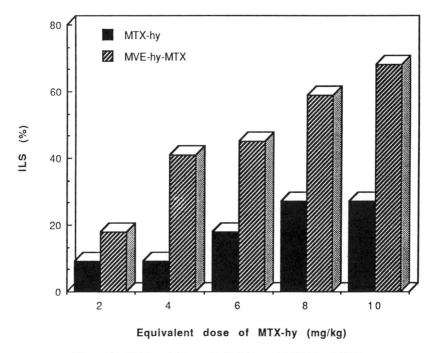

Figure 1. P388 activities of MTX-hy and MVE-hy-MTX.

The polymeric derivative of MTX-γ-hy was tested for its effect against the P388 tumor in a dose range of 2 to 10 mg/kg and showed dose-dependent antitumor activity (Figure 1). MVE copolymer itself, even though it has been reported to have immunostimulating and antitumor activity (*17*), in our hands produced only a 5% ILS when tested at a dose equivalent to that contained in the highest polymeric drug dose. Free MTX-γ-hy was less potent than MVE-γ-hy-MTX at all tested doses. This phenomenon has also been observed with other polymeric drugs such as MVE-5-fluorouridine (MVE-FU) (*5*) and MVE-adriamycin (MVE-Adr) (*6*), although drug potency was usually decreased after covalent linking to polymer.

The effects of MVE-γ-hy-MTX, MTX-γ-hy, MTX and MVE on human leukemia HL60 in athymic mice are shown in Table I. At day 28 post tumor implantation, mice treated with 12 mg/kg of MVE-γ-hy-MTX showed a tumor weight reduction of 96%, compared to the saline control tumor weight, and all animals were still alive (Figure 2). At the same dose, the mean tumor weight of mice treated with MTX-γ-hy was 59% of the saline control. MVE copolymer, at a dose equivalent to the amount in the highest polymeric drug dose (12 mg/kg), produced the tumor which weighed 35% of the saline control. MVE-γ-hy-MTX at 12 mg/kg of MTX equivalents was as active as free MTX at its optimal dose of 1.5 mg/kg in the same test. Figure 3 shows a dose-response graph of MVE-γ-hy-MTX and MTX-γ-hy. For MVE-γ-hy-MTX, a dose-dependent response was observed in the range of 3 to 12 mg/kg.

Table I. In Vivo Activities of MVE-γ-hy-MTX, MTX-γ-hy, MTX and MVE Against Human Leukemia HL60 in Arthymic Mice

Compund	Dose (mg/kg)	%Tumor Weight Compared to Control[a]
Saline (Control)		100
MVE-γ-hy-MTX	12.0	4
	6.0	48
	3.0	78
	1.5	55
MTX-γ-hy	12.0	59
	6.0	81
	3.0	52
	1.5	13
MTX	3.0	22
	1.5	5
	0.75	43
MVE	b	35

[a] At day 28 post tumor implantation.
[b] Dose equivalent to the amount of polymer contained in 12.0 mg/kg of MVE-γ-hy-MTX.

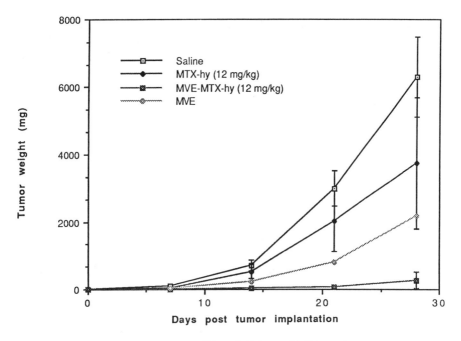

Figure 2. Effects of MVE-hy-MTX, MTX-hy and MVE on human leukemia HL60 in athymic mice.

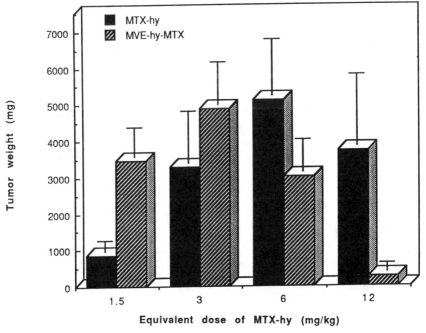

Figure 3. Dose response against human leukemia HL60 in athymic mice.

MTX-γ-hy conjugated to MVE copolymer demonstrated better efficacy and a wider safety margin than MTX-γ-hy. In order to understand its mechanism for the improved activities, biodistribution and pharmacokinetic studies of plasma levels of MVE-hy-MTX need to be carried out.

Acknowledgment

The authors wish to thank Mr. W. A. Hallett for synthesis of methotrexate-γ-hydrazide.

Literature Cited

1. Duncan, R.; Kopecek, J. *Adv. Polymer Sci.* **1984,** *57,* 51.
2. Breslow, D. S. *Pure Appl. Chem.* **1976,** *46,* 103.
3. Butler, G. B. *In Anionic Polymeric Drug Synthesis;* Donaruma, L. G., Ottenbrite, R. M., Vogl, 0. Ed.; John Wiley & Sons: New York, N.Y., 1980.
4. Agency of Industrial Sciences and Technology; Jpn. Kokai Tokkyo Koho JP 60 67,426 1985.
5. Agency of Industrial Sciences and Technology; Jpn. Kokai Tokkyo Koho JP 60 81,197 1985.
6. Agency of Industrial Sciences and Technology; Jpn. Kokai Tokkyo Koho JP 60 67,490 1985.
7. Zunino, F.; Gambetta, R. A.; Penco, S. Belg. BE 902,344, 1985.
8. Zunino, F.; Pratesi, G.; Pezzoni, G. *Cancer Treatment Reports,* **1987,** *7(4),* 367.
9. Agency of Industrial Sciences and Technology; Jpn. Kokai Tokkyo Koho JP 60 67,493 1985.
10. Przybylski, M.; Fell, E.; Ringsdorf, H.; Zaharko, D. *Makromol. Chem.* **1987,** *179,* 1719.
11. Przybylski, M.; Zaharko, D. S.; Chirigos, M. A.; Adamson, R. H.; Schultz, R. M.; Ringsdorf, H. *Cancer Treatment Reports,* **1978,** 62 (11), 1837-1843.
12. Rosowsky, A.; Beardsley, G. P.; Eusminger, W. D.; Lazarus, H.; Yu, C.-S. *J. Med. Chem.* **1978,** *21,* 380.
13. Sirotnak, F. M.; Chello, P. L.; Piper, J. R.; Montgomery, J. A.; De Graw, J. I. *In Chemistry and Biology of Pteridines;* Kisliuk, R. L., Brown, G. M., Ed.; Elsevier/ North-Holland, New York, N.Y., 1979, 597-608.
14. Rosowsky, A.; Forsch, R.; Uren, J.; Wick, M. *J. Med. Chem.* **1981,** *24,* 1450-1455.
15. Rosowsky, A.; Yu, C.-S.; Uren, J.; Lazarus, H.; Wick, M. *J. Med. Chem.* **1981,** *24,* 559.
16. Piper, J. R.; Montgomery, J. A.; Sirotnak, F. M.; Chello, P. L. *J. Med. Chem.* **1982,** *A 25,* 182.
17. Regelson, W. *In Anionic Polymeric Drug;* Donaruma, L. G., Ottenbrite, R. M., Vogl, 0. Ed.; John Wiley & Sons: New York, N.Y., 1980, 303-325.

RECEIVED March 19, 1991

Chapter 10

Succinimidyl Carbonates of Polyethylene Glycol
Useful Reactive Polymers for Preparation of Protein Conjugates

Samuel Zalipsky, Robert Seltzer, and Kwang Nho

Enzon, Inc., 300C Corporate Court, South Plainfield, NJ 07080

A new type of functionalized derivatives of polyethylene glycol (PEG), namely succinimidyl carbonates (SC), were prepared and evaluated as reagents for preparation of PEG-protein conjugates. SC-activated PEGs showed high reactivity towards amino groups of lysine residues producing under mild conditions (pH 7.0 - 10.0, 25 °C, 30 min) extensively modified proteins, in which PEG chains are linked to a polypeptide core through stable urethane (carbamate) linkages. A variety of proteins: chymotrypsin, trypsin, adenosine deaminase, asparaginase, arginase, hemoglobin were subjected to modifications with SC-PEG yielding conjugates with excellent preservation of biological/enzymatic activities.

The unique properties of polyethylene glycol (PEG) and its general biocompatability are the main reasons for the extensive use of this polymer as a modifier of biologically active molecules. These properties include (*1-3*): a wide range of solubilities, lack of toxicity, absence of antigenicity and immunogenicity, non-interference with enzymatic activities and conformations of polypeptides, nonbiodegradability, and ease of excretion from living organisms. Often functionalization of the polymer is the first essential step in preparation of PEG-bioconjugates (*3*). Various functionalized polyethylene glycols have been used effectively in such fields as protein modification (*4,5*), peptide chemistry (*6,7*) and synthesis of polymeric drugs (*8,9*). PEG-protein conjugates useful in medical applications have shown promise, particularly with regard to their stability to proteolytic digestion, reduced immunogenicity and longer half-life times in the bloodstream (*4*).

NOTE: Patent pending.

0097–6156/91/0469–0091$06.00/0
© 1991 American Chemical Society

Originally, for preparation of such conjugates the hydroxyl groups of monomethoxy-PEG (mPEG) were activated with cyanuric chloride, and the resulting compound then coupled with proteins (*10*). This approach suffers from disadvantages, such as the toxicity of cyanuric chloride and its limited applicability for modification of proteins having essential cysteine or tyrosine residues, as manifested by their loss of activity.

Currently, a common form of activated mPEG used for preparation of therapeutic enzymes is mPEG-succinate-*N*-hydroxysuccinimide ester (SS-PEG) (*11*). It reacts with proteins in short periods of time under mild conditions, producing extensively modified conjugates with well preserved biological activity. However, the ester linkage between the polymer and the succinic acid residue has limited stability in aqueous media (*5,12*).

Urethane linkages between amino groups of a protein and PEG provide a stable attachment, more resistant to hydrolytic cleavage (*13*). In fact, it was demonstrated on radioactively labeled PEG-derivatives that urethane links are completely stable under a variety of physiological conditions (*14*). The attachment of PEG to a protein via carbamate was obtained (*15,16*) using carbonyldiimidazole activated PEG. However, the polymer activated in this manner is not very reactive and therefore very long reaction times (48-72 h at pH 8.5) were required to achieve sufficient modifications.

Use of PEG-phenylcarbonate derivatives for preparation of urethane-linked PEG-proteins was reported (*13*). The main drawback of this approach lies in the toxicity of hydrophobic phenol residues (*p*-nitrophenol or 2,4,5-trichlorophenol) and their affinity to proteins.

The present report describes preparation and use of new protein modifying reagents, namely methoxypoly(ethylene glycol)-*N*-succinimidyl carbonate (SC-PEG) and its bifunctional analog, poly(ethylene glycol)-bis-*N*-succinimidyl carbonate (BSC-PEG).

Experimental

Preparation of SC-PEG. *Warning!* Synthesis of this reagent involves the use of phosgene, a highly toxic substance. All manipulations should therefore be performed in a well-ventilated hood.

Methoxypolyethylene glycol of molecular weight 5000 (Union Carbide, 60 g, 12 mmol), dried by azeotropic removal of toluene, was dissolved in toluene/dichloromethane (3:1, 200 mL) and treated with a toluene solution of phosgene (30 mL, 57 mmol) overnight. The solution was evaporated to dryness and the remainder of phosgene was removed under vacuum. The residue was redissolved in toluene/dichloromethane (2:1, 150 mL) and treated with solid *N*-hydroxysuccinimide (2.1 g, 18 mmol) followed by triethylamine (1.7 mL, 12 mmol). After 3 h the solution was filtered and evaporated to dryness. The residue was dissolved in warm (50 °C) ethyl acetate (600 mL), filtered from a trace of insolubles and cooled to facilitate precipitation of the polymer. The product was collected by filtration and then recrystallized once more from ethyl acetate. The product was dried in vacuo over P_2O_5. The yield was 52.5 g (85%).

I.R.(film on NaCl, cm^{-1}) characteristic bands at : 1812 and 1789 (both C=O, succinimide); 1742 (C=O, carbonate); 1114 (CH$_2$OCH$_2$). ^{13}C-NMR (CDCl$_3$): δ 168.5 (CH$_2$C=O); 151.3 (O-CO$_2$); 71.9 (CH$_3$OCH$_2$); 70.2 (PEG); 68.7 (CH$_2$CH$_2$OCO$_2$); 68.0 (CH$_2$CH$_2$OCO$_2$); 58.9 (CH$_3$O); 25.2 (CH$_2$C=O) ppm.

Other SC- and BSC-PEG derivatives were prepared using the same general protocol. The active carbonate contents of the products were determined according to Kalir et al. (*17*) by reacting aliquots of the polymers with an excess of benzylamine and back titration of the latter with perchloric acid in dioxane using thymol blue as an indicator. The results of these determinations are summarized in Table I.

Preparations of PEG-modified proteins. A. SC-PEG (1 g, ≈ 0.2 mmol) was added to a stirred solution of Bovine Serum Albumin (BSA) (100 mg, ≈1.5 x 10^{-6} mol) in 0.1 M sodium phosphate, pH 7.8 (60 mL). Sodium hydroxide (0.5 N) was used to maintain pH 7.8 for 30 min. The excess of free PEG was removed by diafiltration using 50 mM phosphate buffered saline. Approximately 30 amino groups of the native protein were modified as determined by trinitrobenzenesulfonate (TNBS) assay (*28*). The same degree of modification was obtained when the experiment was repeated under identical conditions using SS-PEG instead of SC-PEG.

B. SC-PEG (1 g, ≈ 0.2 mmol) was added to a stirred solution of BSA (100 mg, ≈ 1.5 x 10^{-6} mol) in 0.1 M sodium borate, pH 9.2. Sodium hydroxide (0.5 N) was used to maintain pH 9.2 for 30 min. The excess of free PEG was removed by diafiltration and the product assayed for the number of free amino groups. Approximately 41 of the amino groups of the native BSA were modified.

C. BSC-PEG (1 g, ≈ 0.4 mequiv. SC-groups) was added to a stirred solution of BSA (100 mg, ≈ 1.5 x 10^{-6} mol) in 0.1 M sodium borate, pH 9.2. Sodium hydroxide (0.5 N) was used to maintain pH 9.2 for 30 min. The excess of free PEG was removed by diafiltration and the product assayed for the number of free amino groups. Approximately 48 of the amino groups of the native BSA were modified. Analysis of the product by HPLC (Gel Filtration) indicated that over 65% of PEG-BSA was in an intermolecularly crosslinked form and about 35% of the product had the same molecular weight as PEG-BSA from the above example.

D. Similar methods were used for modification of the enzymes listed in Table II as well as bovine hemoglobin (see Table III). The choice of conditions for the modification reactions (pH, temperature, etc.) was made mainly based on the properties/stability of each protein. Enzymatic activities were measured by previously reported methods (*11, 21-25*).

Determination of Number of PEG Chains Linked to a Protein Molecule. This parameter was calculated by multiplying total number of amino groups on a particular protein by the fraction of amines that were consumed by the modification, which was measured by TNBS assay (*28*) or in some instances by fluorescamine assay (*29*).

Table I. Active Carbonate Content (mole/g x 10^4)
of SC-activated PEG Derivatives

MW of PEG	SC-PEG		BSC-PEG	
	Calculated	Found	Calculated	Found
2000	4.67	4.66	8.76	8.60
3000	3.18	3.33		
4000	2.41	2.37		
4600			4.10	4.02
5000	1.95	1.97		
6000			3.18	3.06

Table II. Properties of SC-PEG Modified Enzymes

Enzyme modified w. SC-PEG (5000)	No. of mPEGs linked per protein molecule	% of native activity (Substrate[*])
Trypsin	7	95(BAEE), 224(ZAPA)
	12	92(BAEE), 326(ZAPA)
Chymotrypsin	9	131(BTEE), 151(BTNA)
	14	122(BTEE), 161(BTNA)
Asparaginase	62	54 (L-Asparagine)
ADA[*]	14	51 (Adenosine)
Arginase	56	71 (L-Arginine)

[*]   Abbreviations: ADA, adenosine deaminse; BAEE, N,α-benzoyl-L-arginine
ethyl ester; BTEE, N,α-benzoyl-L-tyrosine ethyl ester: BTNA, N,α-benzoyl-L-
tyrosine-p-nitroanilide; ZAPA, N,α-benzoylcarbonyl-L-arginine-p-nitroanilide.

Determination of Optimal pH for SC-PEG Reactivity. To triethanolamine-borate buffer (0.3 M, 1 mL) at the appropriate pH, a stock solution of N,α-acetyl-lysine (NAL) in water (50 mM, 0.1 mL) was added followed by a stock solution of SC-activated mPEG-5000 in CH_3CN (50 mM active acyl, 0.1 mL). The resultant solution was vortexed and incubated at 28 °C for 1 h. A mixture of the same components but leaving out SC-PEG was used as a control. The TNBS assay version of Snyder and Sobocinski (*18*) was used to determine the unreacted NAL.

Results and Discussion

We developed an efficient one-pot procedure for preparation of SC-activated PEG (Scheme I). In the first step PEG-chloroformate was generated in situ by treatment of the polymer with phosgene. Then, the polymeric chloroformate was reacted with *N*-hydroxysuccinimide to yield the desired activated PEG derivative. Activated PEG preparations of different molecular weights were obtained this way. They were purified from low molecular weight materials and characterized by infrared and ^{13}C-NMR spectroscopies as well as by titration of the reactive end-groups (*17*). The results of the latter mode of analysis, summarized in Table I, show that the transformation of the functional groups in the process employed proceeded almost quantitatively. The succinimidyl carbonates exhibited excellent storage stability at 4 °C, but the content of the reactive groups decreased slowly at room temperature. For example, SC-PEG derived from mPEG-5000 showed a 15% loss of active acyl groups after 9 months of storage at 25 °C.

As in the case of above-mentioned carbonyldiimidazole activated PEG and PEG-phenylcarbonates, the product of protein modification using SC-PEG or BSC-PEG has PEG-chains grafted onto the polypeptide backbone through carbamate (urethane) linkages (see Scheme II). However, due to the higher reactivity of the new agents, higher degrees of modification are achievable in shorter periods of time. The reactivity of SC-PEG was found to be comparable to the commonly used SS- PEG. For example, under identical conditions (pH 7.8, 25 °C, using 10-fold weight excess of an activated mPEG-5000) both reagents produced PEG-modified bovine serum albumin (BSA) with the same number of substituted amino groups. Kinetic measurements (Zalipsky *et al.*, in preparation) of hydrolysis of both activated polymers and their aminolysis, using N,α-acetyl-lysine (NAL) as a model for the ε-amino group of a protein, showed that SC-PEG has slightly lower reactivity. To evaluate the reactivity of SC-PEG as a function of pH we reacted the activated mPEG-5000 with an equimolar amount of NAL at different conditions, and measured the unreacted amine using trinitrobenzenesulfonate (TNBS) assay (*18*). As shown in Figure 1 the optimal pH for use of SC-PEG was found to be ca. 9.3. The results obtained from actual protein modifications were consistent with this model study. For example, a pH change from 7.8 to 9.2 for the reaction of SC-PEG with BSA, resulted in an increase of total number of mPEG chains linked per protein molecule from 30 to 41. Protein modification reactions were performed in a broad pH range from 6.0 - 10.0 and usually the choice of conditions was based on the properties of the protein involved. All modified proteins were obtained in mild conditions, within

Table III. Properties of PEG - Modified Bovine Hemoglobin

Type of PEG (MW)	No. of PEGs on bHb	P_{50} (mmHg)	1/2 life (h) in rats	%MetHb at 1/2 life
1.	0	28	<1.0	16.5
2. SC-PEG(5000)	3	28	5.5	ND*
3. SC-PEG(5000)	6	28	8.9	ND
4. SC-PEG(5000)	8	24	9.6	5.8
5. SC-PEG(3000)	15	24	12.4	13.8
6. SC-PEG(2000)	24	22	13.0	18.0
7. BSC-PEG(4600)	2	35	ND	ND
8. BSC-PEG(4600) +SC-PEG(5000)	2 3	28	18.8	ND

*ND=not determined

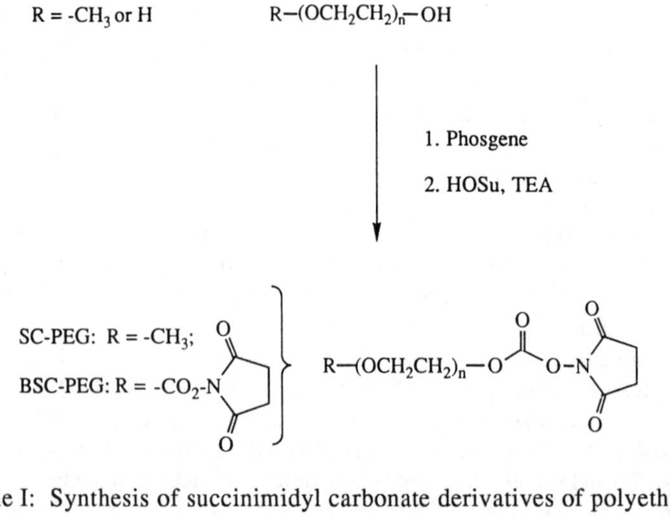

R = -CH₃ or H   R—(OCH₂CH₂)ₙ—OH

1. Phosgene
2. HOSu, TEA

SC-PEG: R = -CH₃;
BSC-PEG: R = -CO₂-N   R—(OCH₂CH₂)ₙ—O—C(=O)—O-N(succinimidyl)

Scheme I: Synthesis of succinimidyl carbonate derivatives of polyethylene glycol.

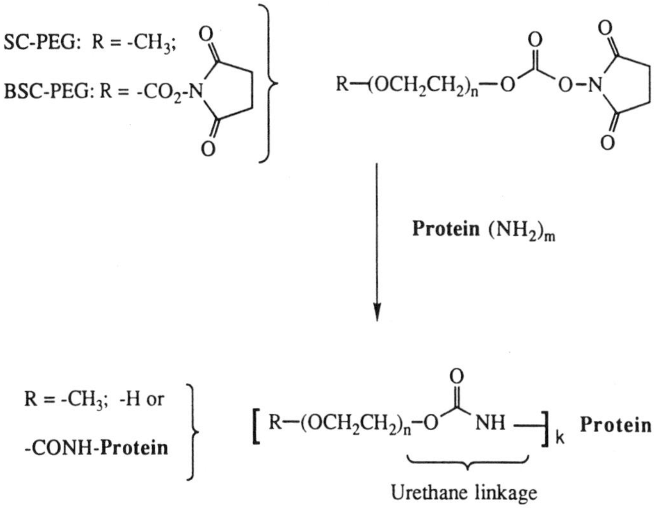

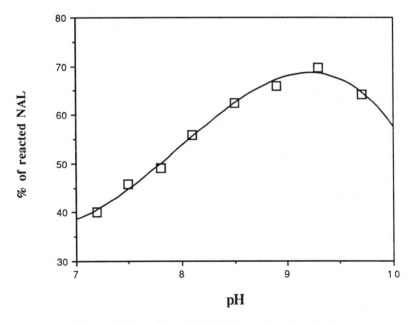

Scheme II: Use of SC-PEG and BSC-PEG for covalent attachment of the polymer to proteins.

Figure 1. Reactivity of SC-PEG as a function of pH.

short periods of time ($\approx$ 30 min), at moderate temperatures (4 - 37 °C). Generally, an SC-activated polymer was added as a powder, which dissolved instantly, to a stirred solution of a polypeptide in an appropriate aqueous buffer. Also the PEG-reagents are soluble in a variety of organic solvents which is useful for coupling of low molecular weight, partially protected peptides and other biologically relevant ligands

The N-hydroxysuccinimide (HOSu) released during polypeptide modification is a nontoxic material that is often used in bioconjugate chemistry as a leaving group-residue of protein modifying reagents. In our experience, unlike hydrophobic aromatic groups (e.g. nitrophenol, trichlorophenol, 13), HOSu does not show affinity towards proteins and can be readily removed from the reaction solutions by dialysis or diafiltration. An additional advantage of SC-activated PEG is that those reactive functional groups that do not react with amino groups of a protein undergo hydrolysis producing HOSu, carbon dioxide and hydroxy-terminated polymer. This feature is of particular importance in the case of the bifunctional BSC-PEG, which can serve dual purpose: PEGylation and crosslinking at the same time. When a BSC-PEG molecule reacts with a protein via only one end group, the other SC-group of the same molecule is hydrolyzed and therefore no extraneous (potentially antigenic) residues are introduced onto the PEG-protein conjugate. Naturally, BSC-PEG under appropriate conditions can be used as a homobifunctional crosslinker of two different proteins. It is also useful as a reactive macromonomer in polycondensation with diamines (19).

PEG Conjugates with Enzymes. The biological activities of enzymes modified with SC-PEG were preserved to a great extent as can be seen from the examples summarized in Table II. In the case of proteolytic enzymes: trypsin and chymotrypsin we actually observed an increase in enzymatic activity of the conjugates towards several low molecular weight substrates. However, these two PEG-enzymes lost a great deal of their proteolytic activity. For example, the mPEG-trypsin adducts were essentially lacking (< 1% of native) proteolytic activity as determined by azocoll assay (20). Both trypsin and chymotrypsin were previously modified with PEG(21-23). The activities of SC-PEG derived conjugates reported here compare most favorably with the literature precedents.

Asparaginase, adenosine deaminase (ADA) and arginase are very important agents for enzyme replacement and antitumor therapy (4). The first two enzymes were reported to lose their activity substantially upon modification with cyanuric chloride-activated mPEG (11,24). In our hands both asparaginase and ADA showed good preservation of activity, even after extensive modifications with SC-PEG. The number of modified sites and the enzymatic activities of the conjugates of these enzymes listed in Table II, are essentially the same as the ones obtained using SS-PEG (11). The SC-PEG - derived modified beef liver arginase exhibited higher enzymatic activity than the previously reported cyanuric chloride-PEG - modified enzyme (25), which had 65% of the native activity even though only 47 mPEG residues were attached to the protein molecule. Visco et al. (26) prepared PEG-arginase in which the polymer strands were bound to the enzyme through urethane linkages. According to this report, the native enzyme was reacted with mPEG-2,4,5-trichlorophenyl carbonate at pH 8.8 for 24 h to produce conjugate in which 90% of the original arginase activity was preserved after coupling of ≈ 53 mPEG-5000 chains per protein. Unfortunately, due to the differences in analytical methods used to characterize this conjugate, reliable comparison to our results is hardly possible.

PEG conjugates with Bovine Hemoglobin. With the goal to develop a prospective blood substitute, SC-PEG and BSC-PEG of various molecular weights were used to modify bovine hemoglobin (bHb). The properties of the PEG-modified hemoglobin derivatives are summarized in Table III. Two types of conjugates were prepared. First, tetrameric, PEG-bHb was derived solely from reactions with SC-PEG (entries 2 - 6). Second, crosslinked (and/or polymerized) PEG-bHb was obtained by use of BSC-PEG (entry 7). The second type of PEG-bHb can be further modified with SC-PEG (entry 8). The P_{50} values (oxygen partial pressure at 50% saturation) of uncrosslinked conjugates remained close to 28 mmHg, of the native bHb. Crosslinking with BSC-PEG produced PEG-bHb with increased P_{50} = 35 mmHg. Higher extent of modification and increased molecular weight of conjugates yielded longer circulating times in rats with little sacrifice in oxygen delivery (compare entries 2-6). This parameter could be extended even further by crosslinking the protein (entry 8). The tendency to form toxic methemoglobin (MetHb) was generally lower in the conjugates than in the native protein, with the exception of the adduct obtained by extensive modification of bHb with mPEG-2000 (entry 6). These results indicate that modification of hemoglobin with SC-activated PEGs allows favorable alteration in physiological and pharmacological properties of the protein while preserving the intrinsic properties of the oxygen-binding site. Overall our results compare favorably with the published data on PEG-Hb conjugates (*27*).

Conclusions

SC-activated PEG derivatives proved to be efficient modifiers of polypeptide materials. These polymeric reagents can be prepared by a simple one-pot procedure from readily available hydroxyl-terminated PEGs. They are sufficiently reactive reagents to produce PEG-protein conjugates, in which the polymer chains are linked to a protein via stable urethane linkages, under mild conditions in short periods of time. The biological/enzymatic activities of such conjugates derived from a number of proteins were preserved to a great extent and compared most favorably to the corresponding PEG-proteins obtained using different coupling chemistries.

Literature Cited

1.   Bailey, F.E.Jr.; Koleske, J.V. *Poly(ethylene oxide);* Academic: New York, NY, 1976.
2.   Merrill, E.; Salzman, E.W. *Am. Soc. Artif. Intern. Organs J.* **1983**, *6,* 60.
3.   Harris, J.M. *J. Macromol. Sci. Rev. Polym. Phys. Chem.* **1985**, *C25,* 325.
4.   Abuchowski, A.; Davis, F. in *Enzymes as Drugs;* Holsenberg, J., Roberts, J., eds.; John Wiley & Sons: New York, NY, 1981, pp 367-383.

5. Dreborg, S.; Akerblom, E.B. *Crit. Rev. Therap. Drug Carrier Syst.* **1990**, *6*, 315.
6. Mutter, M; Bayer, E., in *The Peptides;* Gross, E; Meienhofer, J. Eds.; Academic: New York, NY, Vol. 2; pp 285-332.
7. Zalipsky, S.; Albercio, F.; Slomczynska, U.; Barany, G. *Int. J. Peptide Protein Res.* **1987**, *30*, 740.
8. Zalipsky, S.; Gilon, C.; Zilkha, A. *Eur. Polym. J.* **1983**, *19*, 1177.
9. Ouchi, T.; Yujima, H.; Vogl, O. *J. Macromol. Sci. -Chem.* **1987**, *A24*, 1011.
10. Abuchowski, A.; Van Es, T.; Palczuk, N.C.; Davis, F.F. *J. Biol. Chem.* **1977**, *252*, 3578.
11. Abuchowski, A.; Kazo, G.; Verhoest, C.; Van Es, T.; Kafkewitz, D.; Nucci, M.; Viau, A.; Davis, F. *Cancer Biochem. Biophys.* **1984**, *7*, 175.
12. Ulbrich, K.; Strohalm, J.; Kopecek, *J. Makromol. Chem.* **1986**, *187*, 1131.
13. Veronese, F.; Largajolli, R.; Boccu, E.; Benassi, C.; Schiavon, O. *Appl. Biochem. Biotechnol.* **1985**, *11*, 141.
14. Larwood, D.; Szoka, F. *J. Labelled Compounds Radiopharm.* **1984**, *21*, 603.
15. Beauchamp, C.; Gonias, S.; Menapace, D.; Pizzo, S. *Anal. Biochem.* **1983**, *131*, 25.
16. Berger, H.; Pizzo, S. *Blood* **1988**, *71*, 1641.
17. Kalir, R.; Fridkin, M.; Patchornik, A. *Eur. J. Biochem.* **1974**, *42*, 151.
18. Snyder, S.L.; Sobocinski, P.Z. *Anal. Biochem.* **1975**, *64*, 284.
19. Nathan, A.; Zalipsky, S.; Kohn, J. *Polym. Prepr., Am. Chem. Soc. Div. Polym. Chem.* **1990**, *31(2)*, 213.
20. Chavira, R.Jr.; Burnett, T.J.; Hageman, J.H. *Anal. Biochem.* **1984**, *136*, 446.
21. Abuchowski, A.; Davis, F.F. *Biochem. Biophys. Acta* **1979**, *578*, 41.
22. Pina, C.; Clark, D.; Blanch, H. *Biotechnol. Techniques* **1989**, *3*, 333.
23. Babonneau, M.-T.; Jacquier, R.; Lazaro, R.; Viallefont, P. *Tetrahedron Lett.* **1989**, *30*, 2787.
24. Davis, S.; Abuchowski, A.; Park, Y.K.; Davis, F. *Clin. Exp. Immunol.* **1981**, *46*, 649.
25. Savoca, K.V.; Abuchowski, A.; Van Es, T.; Davis, F.F. *Biochem. Biophys. Acta* **1979**, *578*, 47.
26. Visco, C.; Benassi, C.A.; Veronese, F.M. *Il Farmaco* **1987**, *42*, 549.
27. Yabuki, A.; Yamaji, K.; Ohki, H.; Iwashita, Y. *Transfusion* **1990**, *30*, 516.
28. Habeeb, A. *Anal. Biochem.* **1966**, *14*, 328.
29. Stocks, S.J.; Jones, A.J.M.; Ramey, C.W.; Brooks, D.E. *Anal. Biochem.* **1986**, *154*, 232.

RECEIVED March 19, 1991

Chapter 11

Poly(α-amino acid)–Drug Conjugates
A Biodegradable Injectable Drug Delivery System

Xiaoling Li, David B. Bennett, Nathan W. Adams, and Sung Wan Kim[1]

Center for Controlled Chemical Delivery, Department of Pharmaceutics,
University of Utah, Salt Lake City, UT 84112

Bioactive agents, an opium antagonist and two antihypertensive drugs, have been covalently coupled to biodegradable polymers. Carbamate and carbonate bonds were employed as labile linkages between drug and polymer backbone. In vitro release studies for poly(hydroxypropylglutamine) (PHPG)-prazosin conjugate particles gave 3 weeks constant release and poly(hydroxypropylglutamine-co-leucine)[P(HPG/LEU)]-naltrexone conjugates with various particle sizes showed that the release rate increases with decreasing particle size. In vivo studies were accomplished in male New Zealand white rabbits and female Sprague-Dawley rats by injection of the poly(α-amino acid)-drug conjugate subcutaneously. Following an initial burst, nearly constant drug plasma levels, above minimum effective plasma levels, were achieved for two weeks for the PHPG-prazosin conjugate and 30 days for P(HPG/LEU)-naltrexone conjugates.

Controlled release systems have been used in pharmaceutics and agriculture to deliver bioactive agents at a predicted rate or to a desired site. Polymers have been used as either physical or chemical carriers for bioactive agents. One of the drawbacks for use of inert polymers as drug carriers in implantable devices is that surgical removal is necessary after the drug has been depleted. Biodegradable polymers possess the features of nondegradable polymers with the unique property of being degradable. For systemic drug delivery, biodegradable polymers can be used for subcutaneously implantable, injectable, or insertable devices. Biodegradable polymers include poly(esters), poly(ortho esters), poly(anhydrides), pseudo-poly(amino acid) and poly(amino acids). These can decompose into nontoxic components which are then absorbed or eliminated by the body. Surgical removal of the depleted device is then no longer necessary. An ideal controlled release system for systemic delivery is one

---

[1]Corresponding author.

0097–6156/91/0469–0101$06.00/0

which can release drug at a constant rate. The controlled release of a drug from a biodegradable polymer can be achieved by drug diffusion from the polymer matrix, surface erosion of the polymer, a combination of drug diffusion and surface erosion, or the hydrolysis of a labile covalent bond between the drug and the polymer backbone. The last approach uses a polymeric prodrug concept. Controlled release of bioactive agents has been achieved by the development of poly(α-amino acid)-drug conjugates (1,2,3,4). Such a system is described in Figure 1. Bioactive agents can be liberated by the hydrolysis of the labile bond between a low molecular weight drug and the macromolecule which serves as drug carrier if the polymer backbone is degraded slower than cleavage of the labile bond.

In the treatment of chronic diseases, a long term zero order release dosage form is highly desirable as it reduces fluctuations of drug levels, reduces toxicity and increases patient compliance. Problems in the treatment of both hypertension, a lifetime disorder, and opiate addiction are associated with compliance. The goal of this research is to develop a subcutaneously injectable system which can release drug at constant rates over a long period of time.

Prazosin, an antihypertensive drug with post-synaptic α_1-adrenergic blocking activity, has an amino group and naltrexone, a narcotic antagonist, has two hydroxyl groups suitable to be coupled to a polymer. The polymers used in this study, poly(hydroxypropylglutamine) (PHPG) and its leucine copolymers [P(HPG/LEU)], have been studied by several investigators (5,6,7). Their biodegradability has been demonstrated. The injectable system was designed in a particle form able to pass through a syringe needle. It does not need removal after the device is depleted, since the poly(α-amino acid) can be degraded to nontoxic amino acids and absorbed by the body.

Experimental

Chemicals and Reagents. Prazosin was purchased from the U. S. Pharmacopoeial Convention (Rockville, MD). Naltrexone hydrochloride was obtained from Research Triangle Institute, NC. γ-Benzyl-L-glutamate, L-leucine and PHPG (MW 40,000) was purchased from Sigma (St. Louis, MO). Poly(glutamic acid) (PGA) sodium salt was purchased from Sigma and was converted to the acid form by adjusting pH and precipitation in water. Clonidine hydrochloride was purchased from Sigma and converted to the base form. Triphosgene, 4-nitrophenyl chloroformate and silica gel 60 were purchased from Aldrich (Milwaukee, WI). Trimazosin was obtained from Pfizer Laboratories (Groton, CT) as a gift. LyphoMed heparin sodium injection (10,000 USP Unit/mL) was used. All chemicals used were of reagent, spectrometric, or HPLC grade. Reagent grade dimethylformamide and dimethylsulfoxide were distilled under reduced pressure from appropriate desiccants (phosphorous pentoxide or calcium hydride, respectively) immediately before use. Dioxane, tetrahydrofuran, and absolute ethanol were stored over 4A molecular sieves before use. Triethylamine was distilled over calcium hydride and stored over sodium hydroxide until used. All other chemicals were used as received. New Zealand White male rabbits and Sprague-

Dawley rats were housed individually and were allowed to consume food and water ad libitum with 12 hour periods of light and dark daily.

Spectroscopic Analysis. Infrared (IR) spectroscopic analysis was performed on a Beckman Microlab 620 MX computing spectrometer. Samples were cast on a sodium chloride pellet or made into a pellet with potassium bromide. ^{1}H and ^{13}C NMR spectra were obtained using a JEOL HNM-FX 270 MHz Fourier transform NMR spectrometer. Samples were dissolved in deuterium chloroform and chemical shifts were referenced to an internal standard of tetramethylsilane.

High Performance Liquid Chromatographic (HPLC) Analysis. A Waters HPLC system (two Waters 501 pumps, automated gradient controller, 712 WISP, and 745 Data module) with a Shimadzu RF-535 fluorescence detector or a Waters 484 UV detector, and a 0.5 μm filter and a Rainin 30 x 4.6 mm Spheri-5 RP-18 guard column followed by a Waters 30 x 3.9 cm (10 μm particle size) μ-Bondapak C_{18} column was used. The mobile phase consisted of a 45% aqueous solution (composed of 0.25% triethylamine, 0.9% phosphoric acid, and 0.01% sodium octyl sulfate) and 55% methanol for prazosin analysis or 40% aqueous solution and 60% methanol for naltrexone. The flow rate was 1.0 mL/min. Prazosin was measured by a fluorescence detector at 384 nm after excitation at 340 nm (8) and in vitro release samples of naltrexone were analyzed by UV detection at 254 nm.

Gas Chromatography/Mass Spectra Analysis. The analytical method has been described in the literature (9).

Synthesis of random copolymer of hydroxypropylglutamine and leucine. *N*-carboxyanhydrides (NCA) of γ-benzyl-L-glutamate and L-leucine were prepared by a method described in the literature (10), but modified by replacing phosgene with triphosgene (11). Polymerization proceeded at room temperature in dioxane with a ratio of NCA to triethylamine of 35. The polymer was aminolyzed with 3-amino-1-propanol for five days at 60°C to produce a copolymer of hydroxypropylglutamine and leucine. Molecular weight of the polymer was determined by both viscometry and wide-angle light scattering (WALS) analysis, which were 60,300 and 76,100, respectively.

Synthesis of Naltrexone Monoacetates. Naltrexone-3-acetate was synthesized according to a previously reported procedure with modification (2). Naltrexone (3.07 g) was reacted overnight with 1.1 mL of acetic anhydride in 20 mL of THF with 3.76 mL of triethylamine as catalyst at 0-5°C. The reaction solution was concentrated under high vacuum and the semi-solid residue was purified by flash chromatography. The mobile phase was 2% methanol in chloroform and the stationary phase was silica gel 60 (230-400 mesh). The product (3.21 g, 93.3%) was crystallized from methylene chloride/hexane, m.p. 103-105°C; ^{1}H NMR δ, 6.86, 6.82, 6.69, 6.65 (H-1, H-2 AB quartet), 4.69 (H-5), 3.21 (H-9), and 2.31 (3-acetate methyl H); ^{13}C NMR δ, 20.80 (3-acetate methyl C), 168.42 (3-acetate carbonyl C); IR, 1730 cm^{-1} (6-keto), 1770 cm^{-1}

(3-acetate). Naltrexone-14-acetate was synthesized by selective deacylation of naltrexone-3,14-diacetate (12). Naltrexone (1.28 g) was refluxed in 25 mL of freshly distilled acetic anhydride at 135°C for 5 min and the temperature decreased to 130°C for 25 min. After cooling to room temperature, the reaction solution was concentrated by rotary evaporation. The pH of the residue was adjusted to 8.0 with prechilled 5% sodium bicarbonate. The water phase was extracted with methylene chloride (50 mL) twice. Methylene chloride was evaporated and the crystallization from toluene/hexane yielded 1.26 g (78.7%) of naltrexone-3,14-diacetate, m.p. 144.5-146°C; ^{1}H NMR δ, 6.87, 6.83, 6.70, 6.66 (H-1, H-2 AB quartet), 4.69 (H-5), 4.47 (H-9), and 2.31 (3-acetate methyl H), 2.19 (14-acetate methyl H); ^{13}C NMR δ, 20.73 (3-acetate methyl C), 168.27 (3-acetate carbonyl C), 22.14 (14-acetate methyl C), 170.03 (14-acetate carbonyl C),; IR, 1735 cm^{-1} (6-keto, 14-acetate), 1765cm^{-1} (3-acetate). Naltrexone-3,14-diacetate (1.25 g) was selectively deacylated in 4% sulfuric acid (40 mL) overnight, followed by basification and extraction with methylene chloride. After evaporation of methylene chloride, the residue was crystallized from methylene chloride/hexane to give 0.75 g (66.0%) of naltrexone-14-acetate, m.p. 192-194°C; ^{1}H NMR δ, 6.75, 6.71, 6.62, 6.58 (H-1, H-2 AB quartet), 4.68 (H-5), 4.48 (H-9), and 2.21 (14-acetate methyl H); ^{13}C NMR δ, 22.86 (14-acetate methyl C), 170.11 (14-acetate carbonyl C),; IR, 1725 cm^{-1} (6-keto, 14-acetate).

P(HPG/LEU)-Naltrexone Conjugates. Naltrexone monoacetates (0.60 g and 0.50g for 3- and 14-acetate, respectively) in 10 mL of chloroform were added dropwise with stirring into 20 mL of triphosgene (1.0 g) and triethylamine (0.7 mL) in chloroform. After 1 hour, an additional 1.4 mL of triethylamine was added. The reactions proceeded for 4 hours for 14-acetate and 22 hours for 3-acetate and were evaporated to near dryness by a stream of dry nitrogen. P(HPG/LEU)(50/50), 0.50 g, was dissolved in 27 mL of chloroform/DMF mixture (5:4) and added to naltrexone monoacetate chloroformate at once. The mixtures were stirred for 40-48 hours at room temperature. The reaction solutions were added to 1800 mL of ether and the precipitates were filtered and washed with 200 mL of ether. The products were further purified by dissolution in 50 mL of ethanol and precipitation in 1400 mL of ether. The products were filtered and dried in vacuo to constant weight (0.65 g and 0.62 g for P(HPG/LEU)-naltrexone-14-acetate and (HPG/LEU)-naltrexone-3-acetate conjugates, respectively). The synthetic scheme is shown in Figure 2. The extent of drug loading was determined by dissolving the conjugates in 2% methanolic KOH and quantifying the naltrexone by HPLC.

PHPG-Prazosin Conjugate. P(HPG-*p*-Nitrophenyl carbonate) was prepared by reacting hydroxyl groups on PHPG with 4-nitrophenyl-chloroformate (Figure 3). PHPG (1.000 g, 5.370 mequiv of OH), triethylamine (1.12 mL, 8.04 mM), and 4-dimethylaminopyridine (DMAP, 65.6 mg, 0.537 mM) were dissolved in 45 mL of freshly distilled DMF and the solution cooled to 4°C. 4-Nitrophenyl chloroformate (1.624 g, 8.057 mM) was added and stirred at 4°C. After 72 hours the mixture was then added dropwise, with vigorous stirring, to 1000 mL of ether and stirred overnight. The suspension was filtered and the residue Soxhlet-extracted with

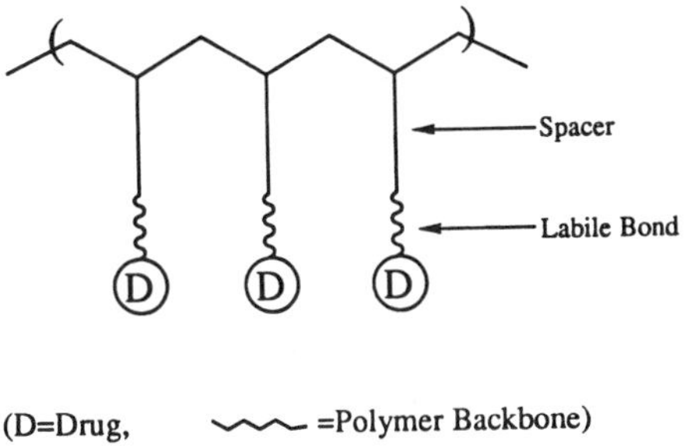

(D=Drug, ∿∿ =Polymer Backbone)

Figure 1. Covalently Coupled Polymer-Drug System.

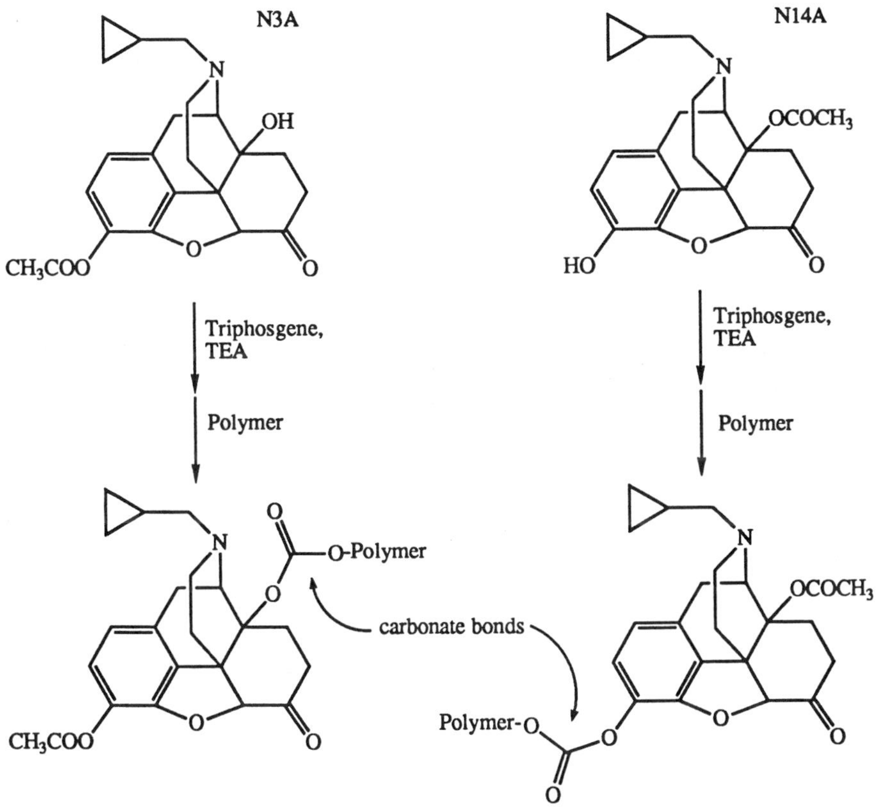

Figure 2. Synthesis of P(HPG/LEU)-Naltrexone Monoacetate Conjugates.

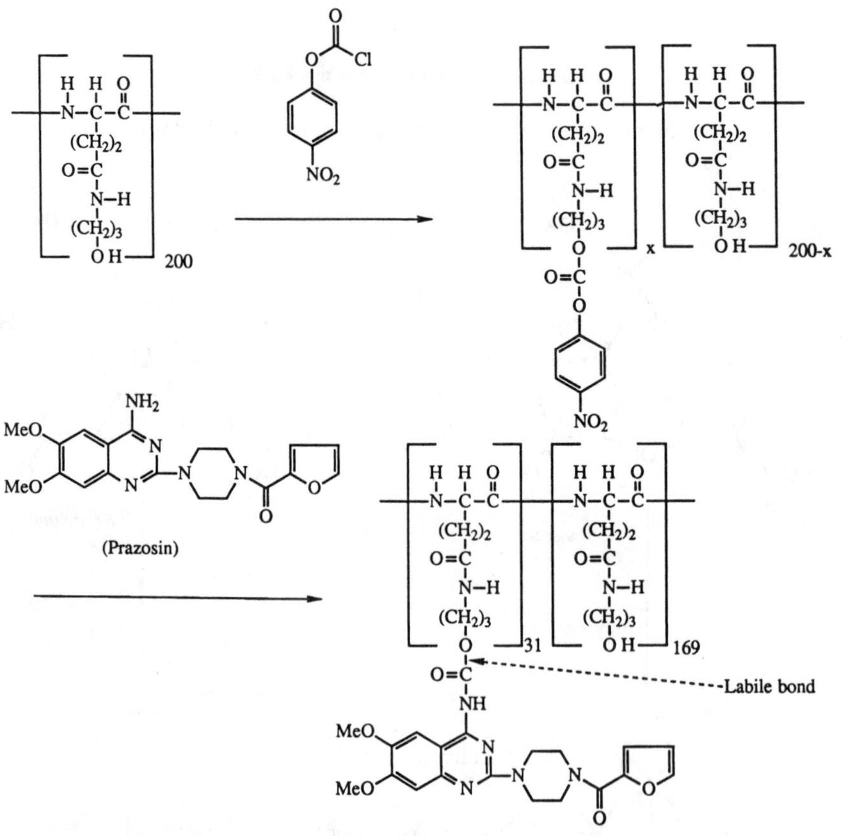

Figure 3. Synthesis of PHPG-Prazosin Conjugate.

absolute ethanol overnight. The residue was then dried under vacuum to give 1.147 g of poly(hydroxypropyl-glutamine-*p*-nitrophenyl carbonate) as an off-white powder; IR: 3450, 3340, 3250, 3020, 2950, 2860, 1750, 1650, 1590, 1570, 1490, 1440, 1390, 1280, 1250, 1220, 1190, 1020, 850, 790 cm^{-1}. P(HPG-Prazosin carbamate) was synthesized by reacting activated polymer with the primary amino group on prazosin(Figure 3). P(HPG-*p*-*N*itrophenyl carbonate) (1.000 g, 2.85 mequiv active ester) was swollen in anhydrous DMSO and sonicated for 20 minutes. Prazosin hydrochloride (1.434 g, 3.42 mM) and triethylamine (0.873 mL, 6.26 mM) were added. The mixture was stirred at room temperature with exclusion of light for 72 hours. The mixture was then poured into 1000 mL of ether and stirred overnight. The suspension was filtered and the residue Soxhlet-extracted with absolute ethanol for three days. The residue was then dried under vacuum to give 1.491 g of product as an off-white powder; IR, 3280, 3065, 2930, 1725, 1712, 1645, 1525, 1440, 1260, 1225, 1050, 870, 790 cm^{-1}. Prazosin loading was determined by accumulating the released free drug from the in vitro release study.

PGA-Clonidine Conjugate. PGA was first reacted with *N*-hydroxysuccinimide to form an activated intermediate and then reacted with clonidine to form the conjugate (Figure 4). PGA (0.50 g) and 0.59 g of *N*-hydroxysuccinimide were dissolved in 30 mL of freshly distilled DMF. Dicyclohexylcarbodiimide (1.05 g) in 5 mL of DMF was dropped into the polymer solution and stirred at room temperature overnight. The reaction solution was kept at -20°C for 4 hours and filtered through a 0.45 μ Teflon membrane to separate the urea formed during reaction. The filtrate was poured into 1000 mL of ether. The white precipitate was collected by filtration and washed with 200 mL of ether. The precipitate was dried under vacuum overnight and 0.65 g of PGA-*N*-hydroxysuccinimide ester was obtained; IR, 3300, 3070, 2960, 1810, 1780, 1740, 1660, 1550, 1430, 1395, 1310, 1260, 1220, 1165, 1085, 1015, 900, 830 cm^{-1}. PGA-*N*-Hydroxysuccinimide ester (0.124 g) was dissolved in 10 mL of DMF and reacted with 0.125 g of clonidine at 80°C for 120 hours. The product was precipitated in 1 liter of ether and washed with acetone, ethanol, phosphate buffered saline (PBS, pH 7.4), and water. Finally, 0.14 g of white solid was obtained after being dried under vacuum, IR, 3300, 3080, 2950, 1670, 1650, 1545, 1450, 1435, 1405, 1350, 1290, 1145, 1100, 1085, 970, 800, 720, 680 cm^{-1}.

In Vitro Release studies. PHPG-Prazosin conjugate, PGA-clonidine conjugate and P(HPG/LEU)-naltrexone conjugates were ground into particles and passed through sieves to obtain various particle size ranges. The particle sizes were verified by microscopy. Particles of the polymer-drug conjugates of known weight and size were placed in dialysis bags (molecular weight cut off 3,500). The bags were filled with 200-250 μl of isotonic PBS (pH 7.4) and placed in 200 mL, 10 mL and 5 mL of PBS receiving fluid for PHPG-prazosin conjugate, PGA-clonidine conjugate and P(HPG/LEU)-naltrexone conjugates, respectively, at 37°C with 50 strokes/min, 2.0 cm/stroke. Release media were replaced daily with fresh PBS upon sampling to maintain sink conditions. Samples were stored at 4°C until analyzed by HPLC.

In Vivo Studies. Male New Zealand white rabbits (2.0 kg) and female Sprague-Dawley rats (180-200 g) were used as animal models for PHPG-prazosin conjugate and P(HPG/LEU)-naltrexone conjugates, respectively. Poly(α-amino acid)-drug conjugates were injected subcutaneously into both flanks of rabbits and the back of rats. PHPG-Prazosin conjugate 100 mg, (20.8 ± 5.6 μm) and 10-32 mg of P(HPG/LEU)-naltrexone conjugates (50-100 μm) were administered as suspensions in saline. Blood samples were collected from the rat tail and rabbit ear veins and centrifuged. The plasma samples were kept at -20°C until analysis. Aliquots of plasma samples were used for analysis. The plasma levels of drugs were determined by HPLC equipped with a fluorescence detector and GC/MS for prazosin and naltrexone, respectively.

Results and Discussion

Controlled release of drug from a covalent bond delivery system usually involves two or three of the following processes: (1) permeation of water into the system, (2) hydrolysis of labile linkage between the drug and polymer backbone, (3) diffusion of free drug from the device to the release medium. Generally, the rate of hydrolysis is slower than the rate of drug diffusion and water permeation. Therefore, drug release can be expected to be controlled by the rate of hydrolysis. For a water insoluble system, the bond activity is considered constant, so a constant release of drug may be approached. There are five major parameters that govern the release of drug from this system: (1) the hydrophobicity and/or solubility of the polymeric prodrug, (2) the length of the spacer group between drug and polymer backbone, (3) the lability of the covalent bond which links the drug to the polymer, (4) drug loading, and (5) the particle size or geometry of the device. The type of covalent bond plays an important role in the controlled release of drugs. The susceptibility of the linkage between drug and polymer backbone to hydrolysis (both enzyme and acid/base catalysis) can be varied by using different types of bonds in combination with the above parameters to control drug release rate. The stability of the various labile bonds is considered to be amide $\geq$ urea > carbamate > carbonate > ester. Esters (*2,3*) and amides (*4*) have been investigated for coupling drugs onto polymers. In this research, carbamate and carbonate bonds have been chosen as the critical linkage. Polymer backbone degradation should be much slower than drug release, since these bonds are less stable than the amide bonds of the polymer backbone.

Our previous study (*2*) with naltrexone revealed that corporation of a hydrophobic component, leucine, into the polymer backbone gave a desired release pattern. P(HPG/LEU) (50/50) was chosen as the backbone to investigate naltrexone delivery because of its hydrophobic nature. After the initial burst, in vitro release rates were almost constant for each particle size range. In vitro release rates of the P(HPG/LEU)-naltrexone-14-acetate conjugate averaged 49.7, 33.9, 30.1, and 24.5 μg/day/100 mg conjugate for particles sized 20-50, 50-100, 100-200, and 200-350 μm, respectively. Release rates of the P(HPG/LEU)-naltrexone-3-acetate conjugate averaged 110.9, 102.5, 87.4, and 79.0 μg/day/100 mg conjugate for particles sized 20-50, 50-100, 100-200, and 200-350 μm, respectively. The release rate increases with decreasing particle size (Figures 5-6). The polymer-naltrexone monoacetate carbonate conjugates had a much lower drug loading (see Table I) than previously studied ester bound conjugates (2) preventing a direct comparison of the effect of type of linkage. The burst observed in vitro may be caused by physical absorption of unbound drug, edge effects, or copolymer chains with higher HPG content which solvate faster and release drug more quickly.

A carbamate linkage was designed to bind prazosin to PHPG. In order to form a carbamate linkage between prazosin and the polymer by using the amino group on prazosin and the pendant hydroxyl group on the side chain of PHPG, the polymer was activated with 4-nitrophenyl chloroformate. Subsequently, an amine can react with the carbonate ester to give a product with a carbamate linkage. IR absorption at 1750 cm^{-1} for activated polymer was an indication of carbonate ester formation. IR absorptions at 1712 and 1725 cm^{-1} indicated formation of carbamate. A low burst showed that most of the free or unreacted prazosin was removed. PHPG-Prazosin conjugate with a particle size of 20.8±5.6 μm gave a constant release for 3 weeks at a rate of 0.92 mg/day/100 mg conjugate (Figure 7). A linear relationship of cumulative release of prazosin vs time (Figure 8) suggests that the release rate was zero order for up to 22 days. Prazosin was the only species detected by HPLC in the release medium. A desired prazosin release rate (1 mg/day) was achieved in the in vitro release study.

Clonidine is a potent drug with a normal dosage of 0.1-0.2 mg/day. A relatively stable linkage or highly hydrophobic polymer backbone should be used to deliver the drug at a low rate. A system with an amide linkage was designed for clonidine by reacting PGA with *N*-hydroxysuccinimide to form an activated intermediate which was then reacted with clonidine. Continuous release of clonidine from the polymeric prodrug was maintained for about 25 days (see Figure 9). A release rate above 0.1 mg/day/20 mg conjugate (particle size 53-90 μm) was observed in an in vitro release study. Only clonidine was detected in the release medium by HPLC.

The rat has been previously used as a model animal for in vivo study of naltrexone (12). As shown in Figures 10 and 11, after an initial burst, the naltrexone plasma levels were 1.03±0.52 ng/mL/mg injected conjugate for P(HPG/LEU)-naltrexone-14-acetate and 0.70±0.39 ng/mL/mg injected conjugate for P(HPG/LEU)-naltrexone-3-acetate for 30 days. From these in vivo data, to achieve a similar plasma level in human, a subcutaneous injection dose of 100 to 500 mg can be predicted assuming that the volume of distribution of naltrexone per kilogram of body weight is of the same magnitude for both rats and humans.

Rabbits have been used in pharmacokinetic and pharmacodynamic studies for prazosin (13,14). The in vivo study with PHPG-prazosin conjugate showed that following the initial burst, nearly constant drug plasma levels were achieved for two weeks at a level of 2.04±0.86 ng/mL/100 mg conjugate injected (Figure 12). The plasma levels were lower than the level predicted by using pharmacokinetic data from literature (13,14) and in vitro release data. This may have been due to the formation of a fibrous capsule around the conjugate particles, which may serve as a diffusional boundary. The plasma levels fell below the detection limit three weeks after

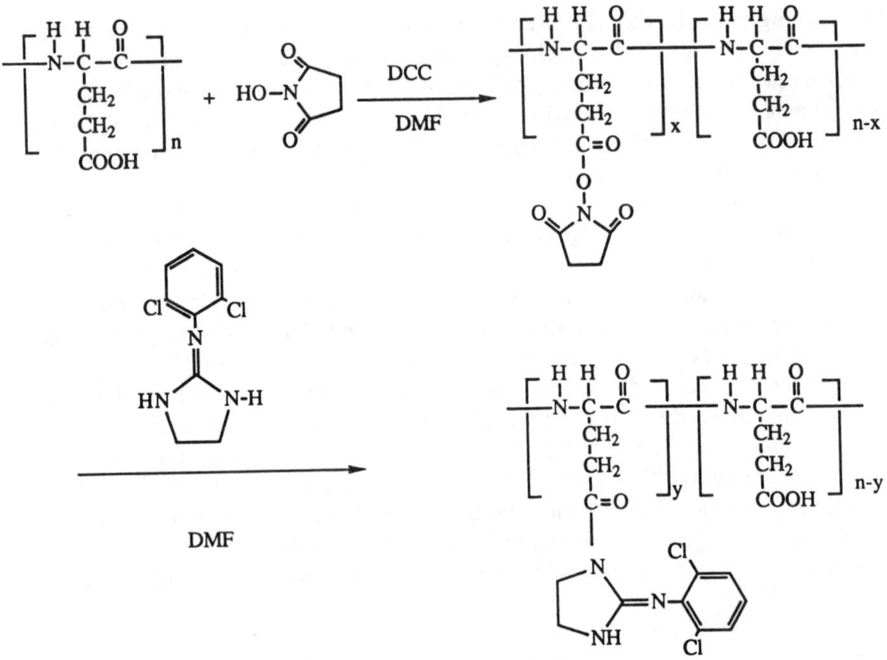

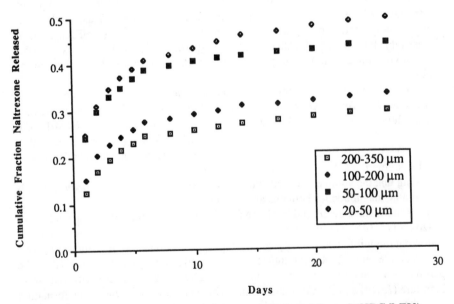

Figure 4. Synthesis of PGA-Clonidine Conjugate.

Figure 5. Cumulative Fraction Naltrexone Release from P(HPG/LEU)-Naltrexone-14-acetate Conjugate Particles of Various Sizes (Loading 8.4%w/w).

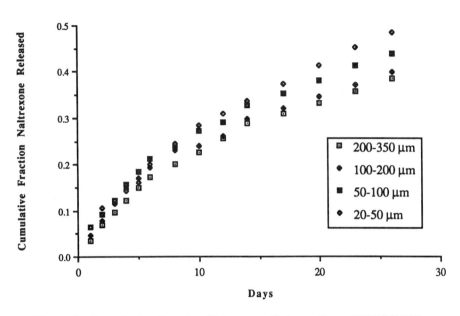

Figure 6. Cumulative Fraction Naltrexone Release from P(HPG/LEU)-Naltrexone-3-acetate Conjugate Particles of Various Sizes (Loading 11.3%w/w).

Table I. Poly(α-amino acid)–Drug Conjugates

Polymer Conjugate	Drug Loading (%w/w)
p(HPG/LEU)-Naltrexone-3-Acetate	11.3
p(HPG/LEU)-Naltrexone-14-Acetate	8.4
PHPG-Prazosin	23.9
PGA-Clonidine	32.8

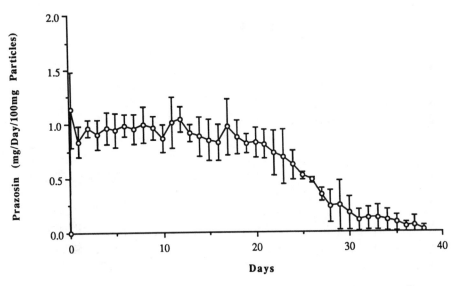

Figure 7. In Vitro Release Profile of PHPG-Prazosin Conjugate (Loading 23.9%w/w).

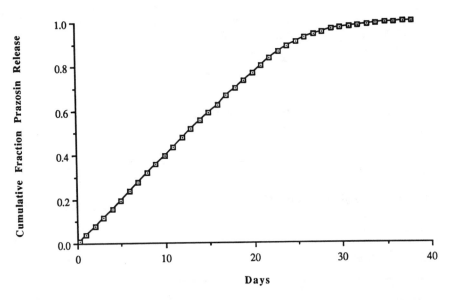

Figure 8. In Vitro Cumulative Fraction Release of Prazosin from PHPG-Prazosin Conjugate.

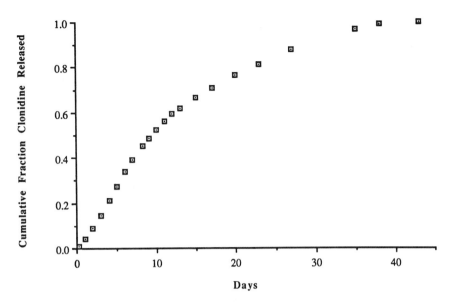

Figure 9. In Vitro Cumulative Fraction Release of Clonidine from PGA-Clonidine Conjugate.

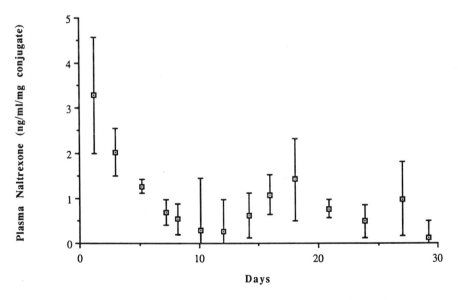

Figure 10. Plasma Levels of Naltrexone Following Injection of P(HPG/LEU)-Naltresone-3-acetate Conjugates (n=3).

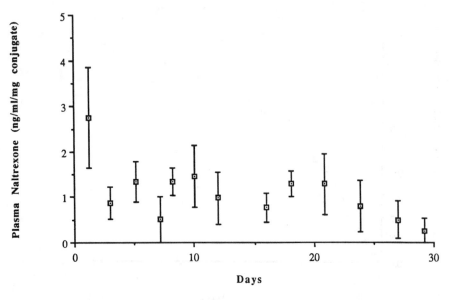

Figure 11. Plasma Levels of Naltrexone Following Injection of P(HPG/LEU)-Naltrexone-14-acetate Conjugates (n=3).

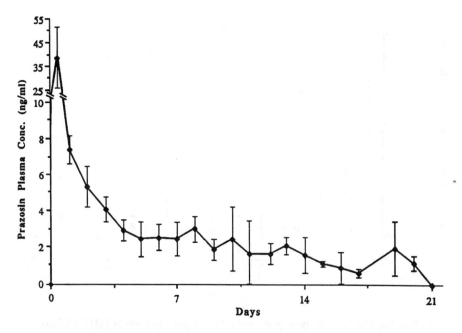

Figure 12. Plasma Levels of Prazosin Following Injection of PHPG-Prazosin Conjugate Subcutaneously in Rabbits (n=3).

injection. This was consistent with in vitro release results which showed a decrease in release rate after 21 days.

An initial burst effect was observed in all in vivo studies. There are several possible factors which may cause a burst effect: physically absorbed free drug, surface effects, and local tissue inflammation during the initial period of injection. It has been shown that inflammation decreases local tissue pH (*15,16*) and causes release of hydrolytic enzymes which would increase the hydrolysis of labile bonds, thereby increasing the release of the drug and, subsequently, increasing plasma levels of drug.

Conclusions

Biodegradable poly(α-amino acids) have been used as chemical carriers to form polymeric prodrugs by covalently binding drug onto polymer via labile linkages. Bioactive agents, an opioid antagonist and two antihypertensive drugs, were delivered by subcutaneous administration. In vitro and in vivo studies for poly(α-amino acid)-drug conjugates showed promising results and may represent a new approach in the treatment of chronic diseases.

Acknowledgments

We would like to thank Dr. J. Feijen and Dr. E. Mack for stimulating discussion. This research was supported, in part, by NIH grant HL44539-04A1.

Literature Cited

1. Feijen, J.; Gregonis, D.; Anderson, C., Petersen, R. V., and Anderson, J. *J. Pharm. Sci.*, **1980**, *69*, 871.
2. Negishi, N.; Bennett, D. B.; Cho, C.-S.; Jeong, S. Y.; Van Heeswijk, W. A. R.; Feijen, J., and Kim, S. W. *Pharm. Res.*, **1987**, *4*, 305.
3. Petersen, R. V.; Anderson, C. G.; Fang, S-M.; Gregonis, D. E.; Kim, S. W.; Feijen, J.; Anderson, J. M. and Mitra, S. In *Controlled Release of Bioactive Materials;* R. Baker, Ed.; Academic Press, New York, 1980, pp.45-61.
4. Bennett, D. B.; Adams, N. W.; Li, X., and Kim, S. W. *J. Bioact. Compat. Polym.*, **1988**, *3*, 44.
5. Hayashi, T.; Tabata, Y., and Nakajima, A. *Polymer J.*, **1985**, *17*, 463.
6. Rypacek, F.; Saudek, V.; Pytela, J.; Skarda, V., and Drobnik, J. *Makromol. Chem., Suppl.*, **1985**, *9*, 129.
7. Minoura, N.; Aiba, S., and Fujiwara, Y. *Makromol. Chem., Rapid Commun.*, **1984**, *5*, 643.
8. Lin, E. T.; Baughman, Jr., R. A.; and Benet, L. Z. *J. Chromatogr.*, **1980**, *183*, 367.
9. Monti, K. M.; Foltz, R. L., and Chin, D. M. *J. Anal. Toxicol.*, Submitted.
10. Fuller, W. D., Verlander, M. S., and Goodman, M. *Biopolymers*, **1976**, *15*, 1869.

11.   Daly, W. H.; Poche, D. *Tetrahedron Letters,* **1988,** *29,* 5859.
12.   Dewey, W. L.; Harris, L S.; Howes, J. F., and Nuite, J. A. *J. Pharmacol. Exp. Ther.,* **1970,** *175,* 435.
13.   Hamilton, C. A.; Reid, J. L., and Vincent, J. Br. *J. Pharmacol.,* **1985,** *86,* 79.
14.   Vincent, J.; Hamilton, C. A., and Reid, J. L. *Clin. Exp. Pharmacol. Physiol.,* **1986,** *13,* 593.
15.   Menkin, V. *Am. J. Pathol.,* **1934,** *10,* 193.
16.   Zaikov, G. E. *J. Macromol. Sci.,* **1985,** *C25,* 551.

RECEIVED March 19, 1991

Chapter 12

Transcellular Transport of Protein–Polymer Conjugates in Cultured Epithelial Cells

Release of a Protein from Its Polymeric Carrier during Transcytosis

Wei-Chiang Shen, Jiansheng Wan, and Mitchell Taub

Drug Targeting Research, Division of Pharmaceutics, School
of Pharmacy, University of Southern California, Los Angeles, CA 90033

Poly(L-lysine) (PLL) and poly(D-lysine) (PDL) were used as models of biodegradable and non-degradable polymeric carriers, respectively, in the studies of transepithelial transport of a protein, horseradish peroxidase (HRP), in cultured Madin-Darby canine kidney (MDCK) cells. Conjugates of HRP to these carriers were prepared using either a thioether (-S-) or a disulfide (-SS-) linkage. Transport of HRP-S-PLL, but not HRP-S-PDL, was found only in the basal-to-apical direction, and was inhibited by leupeptin. No transport of either conjugate was detected in the apical-to-basal direction, indicating that a specific proteolytic compartment exists in the basal-to-apical pathway where the PLL moiety in the conjugates can be degraded and free HRP can be released into the apical medium. On the other hand, the magnitude of HRP-SS-PDL transport in the apical-to-basal direction was found to be significantly higher than that of the basal-to-apical direction. It was also found that the disulfide linkage in the HRP-SS-PDL conjugate could be cleaved on both cell surfaces, and this disulfide-cleavage activity was higher at the basal membrane than at the apical membrane. These results demonstrate that the release of an intact protein from its polymeric carrier during the transcellular transport process is an important factor in determining the efficiency of transcytosis of a protein-polymer conjugate in epithelial cells.

Polymers, being macromolecules of considerable size and weight, have many limitations when used either as drugs or as drug carriers (*1*). One of the most serious limitations is the existence of epithelial or endothelial barriers (*2*). However, macromolecules can be transported by a vesicular process known as transcytosis (*3,4*). In transcytosis, a polymer can be shuttled across an epithelial cell by first

0097–6156/91/0469–0117$06.00/0

binding to a small piece of the plasma membrane and subsequently becoming entrapped inside a small vesicle. These polymer-containing vesicles can be processed within the cell, translocated across the cell cytosol, and then are able to release their contents via fusion with the plasma membrane on the opposite side of the cell monolayer. However, a polymer simply having a high affinity for the plasma membrane is not necessarily a polymer that can be transported across the cell. It is conceivable that once the macromolecule is translocated to the opposite side of the cell it remains attached to the plasma membrane without being released into the medium. Such a translocation without dissociation of membrane-bound macromolecules has been observed in the transcytosis of cationic ferritin from the apical to the basolateral surface in the epithelium of rat choroid plexus (5). Therefore, the dissociation or degradation of the polymer from the plasma membrane becomes an important process to elucidate if transcytosis will be successfully used as a mechanism for the transepithelial or transendothelial delivery of polymers to target tissues.

In this report, we used polylysine as a model polymer for the study of transcytosis (6). Polylysine has several advantages as a polymeric carrier, such as its positive charge and high affinity for the negatively charged plasma membrane. Amino groups on polylysine can be easily modified and conjugated to other large or small molecules. In addition, both D and L forms of polylysine (PDL and PLL) are available; hence, a comparison between nondegradable (PDL) and degradable (PLL) polymers can also be obtained (7).

Experimental Procedures

The Strain 1 MDCK cell line was a gift from Dr. M.J. Cho, the Upjohn Co. Cell culture medium/reagents and trypsin were obtained from Gibco Laboratories (Grand Island, NY), and Transwells (2.45 cm dia., 0.4 μ pore size) were obtained from Costar, Cambridge, MA. Horseradish peroxidase (HRP), poly(D-lysine) HBr, M_r 54,000 (PDL), and Poly(L-lysine) HBr, M_r 52,000 (PLL), were purchased from Sigma Chemical Co., St. Louis, MO. Protein cross-linkages were established using the sulfo-SMPB and SPDP reagents, which are products of Pierce, North Chicago, IL. Cell stocks were grown in Corning's tissue culture flasks in Eagle's minimal essential medium (MEM) supplemented with 10% fetal bovine serum (FBS).

Cell cultures. MDCK cells were seeded in the Transwells at a density of 2.2×10^4 cells/cm^2. Cells were fed by changing medium in both upper (apical) and lower (basal) compartments periodically. Confluent monolayers were obtained at 5-7 days post-inoculation, when the cell density reached $4.5\text{-}5.0 \times 10^5$ cells/cm^2, and a transepithelial electrical resistance (TEER) of about 2,000 ohms·cm^2 was measured using an epithelial voltohmmeter (EVOM, World Precision Instruments, West Haven, CT). The amount of FBS in the cell culture medium could be decreased as the cells approached their maximum resistance, and could be maintained at that point for 2 days or longer in medium containing 1% FBS.

Preparation of HRP-SS-PDL and HRP-S-PLL. HRP was conjugated to PLL via a stable thioether linkage, and was conjugated to PDL via a reducible disulfide linkage as described (8,9). Briefly, for the synthesis of the HRP-S-PLL linkage, HRP (10 mg) was modified with 1.5 mg of sulfo-SMPB in 1 mL of phosphate-buffered saline solution (PBS), pH 7, for 1 h at 25°C. Following extensive dialysis, the modified HRP was mixed with 10 mg of PLL which had been previously modified by SPDP, reduced by dithiothreitol (DTT), and purified by G-50 Sephadex gel filtration. The reaction mixture was left to stir for 16 h, and the conjugate was purified using a Sephadex G-100 column. The HRP and polylysine contents in the conjugate were determined by spectrophotometric measurement at 403nm and the trypan blue precipitation method (8,9), respectively. For synthesis of the HRP-SS-PDL linkage, HRP was modified by SPDP (as opposed to sulfo-SMPB) in PBS, pH 7.4 for 1 h at 25°C. Excess SPDP was removed by overnight dialysis at 4°C against the same buffer. SPDP-modified HRP (10 mg) was then added to sulfhydryl-containing PDL (10 mg), and after mixing for 16 h at 4°C, the conjugate was purified using Sephadex G-100 gel filtration. The molar ratio of HRP to the poly (amino acid) was approximately 1:1 in both conjugates, and about 60% of the enzyme activity of HRP was retained in both conjugates.

Transcellular transport of HRP conjugates. HRP conjugates were added directly to the apical or basal medium of Transwells having confluent MDCK monolayers, giving a final concentration of 3 μg/mL HRP. At various time intervals during transport experiments, 100 μL aliquots of medium were pipetted from the opposite side of the cells to conjugate administration, and an equal volume of fresh medium was then added as a constant volume replenishment. HRP activity in each of the samples was measured spectrophotometrically using the o-dianisidine colorimetric method (10). In some experiments, ^{14}C-sucrose was added simultaneously to the basal medium at 1 μCi/mL in order to monitor the integrity of the tight junctions. In the study of trypsinized HRP-S-PLL, the conjugate was pretreated with 0.025% trypsin in PBS, pH 7 at 37°C for 5 min prior to addition to the basal side of the Transwell. HRP enzyme activity in the trypsinized HRP-S-PLL was about 20% higher than that of the original conjugate.

Results and Discussion

The in vitro system we have been using to study the transepithelial transport is cultured Madin-Darby canine kidney (MDCK) epithelial cells (11). When cultured on microporous polycarbonate filters (Transwell, Costar, Cambridge, MA), MDCK cells will develop into monolayers mimicking the mucosal epithelium (11). When these cells reach confluence, tight junctions will be established between the cells, and free diffusion of solutes across the cell monolayer will be markedly inhibited. Tight junction formation can be monitored by measuring the transepithelial electrical resistance (TEER) across the cell monolayers. In Figure 1, MDCK cells were seeded at 2 X 10^4 cells per well in Transwells (0.4 μ pore size) as described previously. TEER and ^{14}C-sucrose transport were measured daily. To determine ^{14}C-sucrose transport, 1 μ Ci of the labeled compound was added to the apical medium, and, after 1 h of incubation, 0.1 mL of the basal medium was pipetted and radioactivity levels were determined using a scintillation counter. This value demonstrates the change in TEER and ^{14}C-sucrose diffusion rates across a microporous filter in response to the growth of MDCK cells on this filter. Therefore, after confluence is reached, polymers will be unable to permeate between intercellular junctions in an MDCK cell monolayer.

In this report, we have conjugated an enzyme, horseradish peroxidase (HRP) with either PLL or PDL by two types of linkages, i.e. the disulfide- and thioether-linkages. These two conjugates, HRP-S-PLL and HRP-SS-PDL, are sensitive to proteolysis and reduction, respectively. The release of HRP from its carrier is illustrated in Figure 2, where HRP-S-PLL represents a protein that is conjugated to a biodegradable polymer; hence, free HRP can be released from the conjugate upon the degradation of the PLL moiety. HRP-SS-PDL represents a protein that is conjugated to a nondegradable polymer via a cleavable linkage; hence, HRP can be released from PDL upon the reduction of the disulfide linkage in the conjugate. If, during the transcellular transport process, the HRP-S-PLL conjugate encounters a proteolytic environment/compartment with trypsin-like enzyme activity (8), the PLL moiety will be degraded and free HRP can be released. Also, if during the transcellular transport process the HRP-SS-PDL conjugate encounters a disulfide bond-reducing environment (9,12), the PDL moiety in the conjugate will be separated from the HRP, and again free HRP can be released.

When PLL conjugates were tested in MDCK cell monolayers, two unexpected observations were noted. First, the transport of HRP by the PLL carrier occurred only in the basal-to-apical direction. Secondly, the transport of HRP by the PLL carrier in the basal-to-apical direction proceeded only after a 2-h lag time. As shown in Figure 3, confluent MDCK monolayers in Transwells were treated either at the basal or apical compartment with 3 μ g/mL HRP-S-PLL conjugate. At various time intervals, an aliquot (0.1 mL) of the medium was pipetted from the compartment opposite to conjugate administration and the HRP activity was assayed immediately. After 5 h, the HRP transport in the basal-to-apical direction was at least 10-fold higher than that in the apical-to-basal direction. It is known that, in the fluid-phase transcytosis of HRP, apical-to-basal transport is normally higher compared to that of the basal-to-apical direction (13). In Figure 4, confluent MDCK monolayers in Transwells were treated at either the basal or apical compartment with ^{14}C-sucrose (1 μ Ci/mL). At various time intervals, an aliquot (0.1 mL) of the medium was pipetted from the opposite compartment and the ^{14}C-radioactivity was measured. After 5 h, the ^{14}C-radioactivity level counted in the apical-to-basal direction was approx. 50% higher than that of the basal-to-apical direction. TEER measurements showed a small, yet negligible, decrease in resistance after 5 h. This demonstrates that, although there was no detectable apical-to-basal transport of HRP-S-PLL, the apical-to-basal transport of ^{14}C-sucrose was actually higher than that shown in the basal-to-apical direction. In addition, it is known that the half-life of plasma membrane-bound ligands in transcytosis can be as short as only 30 min (14). Therefore, the 2-h lag

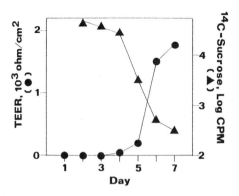

Figure 1. The correlation of transepithelial electrical resistance (TEER) with the transepithelial transport of ^{14}C-sucrose in MDCK cell monolayers grown on microporous filters.

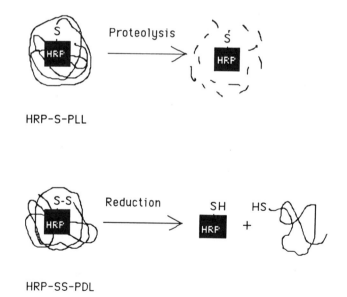

Figure 2. Two mechanisms of releasing a protein from its polymeric conjugates.

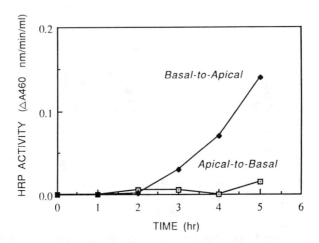

Figure 3. Transcellular transport of HRP-S-PLL in filter-grown MDCK cell monolayers. Confluent MDCK monolayers in Transwells were treated at the basal compartment (closed squares) or the apical compartment (open squares) with 3 µg/mL HRP-S-PLL conjugate.

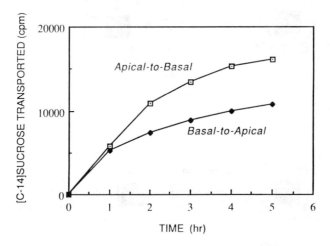

Figure 4. Transcellular transport of ^{14}C-sucrose in filter-grown MDCK cell monolayers. Confluent MDCK monolayers in Transwells were treated at the basal compartment (closed squares) and the apical compartment (open squares) with ^{14}C-sucrose (1 µCi/mL).

time suggests that a slow digestion process occurs during transcytosis of the HRP-S-PLL conjugate.

As shown in Table I, free HRP is poorly transported across MDCK cells; but, when conjugated to a PLL carrier, HRP transport is increased considerably. The existence of a proteolytic compartment involved in the transcytotic digestion of HRP-S-PLL conjugate was further confirmed by the finding that when PLL was replaced by PDL, the transport of HRP was completely abolished (Table I) (*8*). In addition, when protease inhibitors such as leupeptin were added to the basal medium, the transcytosis of HRP was also significantly decreased (Table I). We have previously reported that the partial degradation of HRP-S-PLL was not inhibited by lysosomotropic amines (*8*), indicating that this proteolytic process does not occur in lysosomes.

Our results suggest that in the transcytosis of HRP-S-PLL in the basal-to-apical direction there exists a non-lysosomal compartment where the PLL moiety is degraded and preparation is made for the subsequent release of HRP to the apical medium. This processing of PLL takes approximately 2 hours to complete. In the apical-to-basal direction, however, an absence of this compartment in the pathway renders the cell devoid of a mechanism for the transport of HRP conjugate from the apical to the basal medium. Referring to Figure 5, the postulated routes of intracellular processing of the HRP-S-PLL conjugate are shown. Apically internalized conjugate (A) can either be directed to a lysosomal compartment (B), where it is enzymatically digested and subsequently released as degradation products at both surfaces (C), or, it can be transcytosed as intact conjugate to the basal surface (D). Due to lack of proteolytic enzymes in this pathway, HRP will not be released; rather, it will adhere to the basal membrane for possible basal-to-apical transcytosis (E). Basally internalized conjugate (F) may also be directed to a lysosomal compartment (B); but, it may also cross the cell via a non-lysosomal proteolytic compartment (G) where the PLL moiety can be selectively cleaved. Free and intact HRP (H) can then be released into the apical medium.

When HRP-SS-PDL, a disulfide-linked conjugate, was added to both the apical and basal medium of cultured MDCK cells in Transwells, contradictory results were obtained as compared with those from the use of HRP-S-PLL conjugate. The first difference noted was the lack of a lag time during the transport of HRP-SS-PDL. Furthermore, the long term transport of HRP-SS-PDL in the apical-to-basal direction was significantly higher than that of the basal-to-apical direction, despite an initial surge of basal-to-apical transport which was observed during the first 5 h of incubation. This polarity is illustrated in Figure 6, which shows a 3-fold increase in the transport of membrane bound HRP in the apical-to-basal direction vs. that of the apical-to-basal direction following a 20 h incubation. The absence of an appreciable lag time in the transcytosis of HRP-SS-PDL suggests that it is not subjected to a time-consuming process within its transcytotic pathway. We have compared the activity levels of disulfide cleavage inherent to both cell membrane surfaces in MDCK cells, and have found that the basal surface has a much higher activity than that of the apical surface with respect to disulfide bond cleavage ability (Table II). Also, a large proportion of basal membrane-bound HRP-SS-PDL was released as free HRP back

**Table I. Basal-to-Apical Transcytosis of HRP-S-PDL
in Cultured MDCK Cell Monolayers[a]**

	HRP Transcytosed in 5hr (A460 nm/min/mL)
HRP	0.02
HRP-S-PLL	0.15
+Leupeptin[b]	0.04
HRP-S-PDL	0.02

[a] Experimental procedure as described in text.
[b] Leupeptin was added at 20 μg/mL to the basal medium.

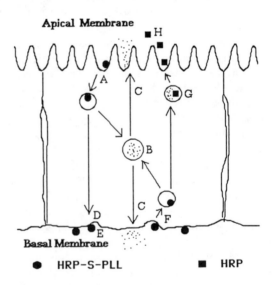

Figure 5. Schematic illustration of the pathway involved in the transcellular transport of HRP-S-PLL.

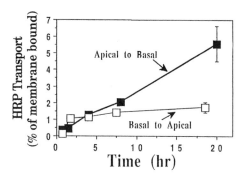

Figure 6. Transcellular transport of HRP-SS-PDL in a filter-grown MDCK cell monolayer. HRP-SS-PDL was added to the apical medium (closed squares) or to the basal medium (open squares).

**Table II. Release of HRP in HRP-SS-PDL and HRP-S-PDL
from the Apical and the Basal Surface of MDCK Cell Monolayers[a]**

| | HRP released in 4hr (% of membrane-bound HRP) | |
	Apical Surface	Basal Surface
HRP-SS-PDL	16.7 ± 1.0	44.6 ± 8.0
HRP-S-PDL	7.4 ± 0.6	7.8 ± 1.4

[a] MDCK cell monolayers were preincubated at 4°C for 30 min with 3 μ g/mL of HRP-SS-PDL or HRP-S-PDL in the bassal or the apical medium. Following the removal of unbound conjugate, the release of HRP was determined by reincubation of the monolayers at 37°C.

into the basal medium instead of the conjugate (*9*). Because the basal membrane-bound HRP-SS-PDL is effectively released from the surface, there is little conjugate available for transport after a prolonged incubation. This finding is consistent with data obtained from the transport studies, where no increase of HRP-SS-PDL transport was detected after 5 h. However, apical membrane-bound HRP-SS-PDL was released at a rate much slower than that of basal membrane-bound conjugate. It took 20 h to reach the same level of HRP activity compared with 5 h for the basal surface (*9*). Conceivably, the slower HRP release rate from apical membrane-bound conjugates would allow more HRP-SS-PDL to be available for apical-to-basal transport. This finding suggests that the magnitude of transcytosis of disulfide-linked conjugates is limited by the disulfide cleavage activity on the binding surface. Figure 7 illustrates the postulated routes of intracellular processing of internalized HRP-SS-PDL conjugate. Apically internalized conjugate (A) can release free, intact HRP via

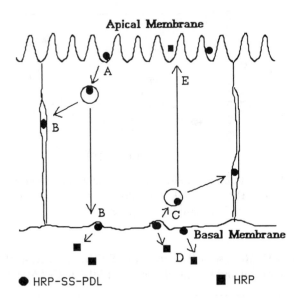

Figure 7. Schematic illustration of the pathway involved in the transcellular transport of HRP-SS-PDL.

transport to the basolateral membrane surface (B), as the basal membrane has been shown to harbor the necessary reductive activity to elicit cleavage of the disulfide linkage (Table II). The apical membrane posesses relatively low reducing activity (compared to that of the basal membrane), and therefore cannot effectively accomplish conjugate cleavage. Basal membrane-bound conjugate (C), however, may have two separate fates. One possibility is the reduction of the disulfide linkage at the basal surface; hence, no internalization of conjugate and release of free, intact HRP into the basal medium (D). The other possibility, which occurs only initially, is the bypassing of the reduction at the basal surface, transcytosis, and subsequent release of free, intact HRP into the apical medium (E). These hypotheses are experimentally qualified from data presented in Figure 6 and Table II.

Results from our studies investigating the transcytosis of different HRP-polylysine conjugates indicate that the release of HRP from epithelial membranes could be a limiting step in the transcellular transport of polymers as drugs or drug-carriers. Such a releasing mechanism has been found in the natural transcytotic process of physiologically important macromolecules. For example, the transcytosis of IgG molecules in epithelial cells of the neonatal intestine is facilitated by the difference in pH between the luminal and basal membranes which can promote dissociation of the IgG from the Fc-receptor (15). On the other hand, transcytosed IgA molecules are released from intestinal epithelial cells by the proteolytic digestion of the IgA-receptors (14).

Conclusions

We show here that biodegradable polymers (PLL) and spacers (disulfide bond) can be used in promoting transcytosis, while non-biodegradable polymers and non-cleavable linkages cannot. Therefore, in designing an effective polymeric carrier for transcellular transport of proteins or drugs, one must consider not only the membrane-binding, membrane-internalization, and membrane-translocation behavior of the epithelial cell, but also the release of the carried protein or drug molecules from their carriers once they are transported to the opposite side of the epithelium.

Acknowledgments

This investigation was supported by Grant PHS CA 34798 from the National Cancer Institue.

Literature Cited

1. Duncan, R.; Kopecek. *J. Adv. Polymer Sci.* **1984,** *57,* 53
2. Poznansky, M.J.; Juliano, R.L. *Pharmacol. Rev.* **1984,** *36,* 277.
3. Mostov, K.E.; Simister, N.E. *Cell* **1985,** *43,* 389.
4. Rodman, J.S.; Mercer, R.W.; Stahl, P.D. *Current Opinion in Cell Biology* **1990,** *2,* 664.
5. van Deurs, B.; von Bulow, F.; Moller, M. *J. Cell Biol.* **1981,** *89,* 131.
6. Ryser, H.J.-P.; Shen, W.-C. In *Targeting of Drugs with Synthetic Systems;* Gregoriadis, G.; Senior, J.; Poste, G., Eds; Plenum Press, New York, NY, 1986; pp. 103-120.
7. Shen, W.-C.; Ryser, H.J.-P. *Mol. Pharmacol.* **1979,** *16,* 614.
8. Shen, W.-C.; Wan, J.; Shen, D. *Biochem. Biophys. Res. Commun.* **1990,** *166,* 316.
9. Wan, J.; Persiani, S.; Shen, W.-C. *J. Cell. Physiol.* **1990,** *145,* 9.
10. *Worthington Enzyme Manual;* Worthington Biochem. Co., Freehold, NJ, 1972; pp. 43-45.
11. Cho, M.J.; Thompson, D.P.; Cramer, C.T.; Vidmer, T.J.; Scieszka, J.F. *Pharm. Res.* **1989,** *6,* 71.
12. Feener, E.P.; Shen, W.-C.; Ryser, H.J.-P. *J. Biol. Chem.* **1990,** *265,* 18780.
13. Bomsel, M.; Prydz, K.; Parton, R.G.; Gruenberg, J.; Simons, K. *J. Cell Biol.* **1989,** *109,* 3243.
14. Mostov, K.E.; Deitcher, D.C. *Cell* **1986,** *46,* 613.
15. Rodewald, R. *J. Cell Biol.* **1980,** *80,* 18.

RECEIVED March 19, 1991

Chapter 13

Opportunities for Protein Delivery by Aerosol
Absorption and Retention of Several Synthetic Polypeptides in the Lung

Peter R. Byron[1], Hirokazu Katayama[1], Zhuang Sun[1],
and Frantisek Rypacek[2]

[1]School of Pharmacy, Virginia Commonwealth University,
Richmond, VA 23298
[2]Institute of Macromolecular Chemistry, Prague, Czechoslovakia

Inhalation offers some exciting possibilities for the delivery of new generation products of biotechnology like polypeptides and proteins. This article reviews some of the possibilities and describes a series of experiments in which the pulmonary absorption of a peptidase-resistant, synthetic polypeptide, poly-α,β-[N(2-hydroxyethyl)-D,L-aspartamide], containing a covalently bound fluorophore (F-PHEA), was studied in the isolated rat lung. Effects of administered dose and polydisperse molecular weight distribution (MWD) are described as they affect the amount of F-PHEA absorbed and its MWD following transfer to the pulmonary vasculature. Relatively small molecules, less than 5 kilo Daltons (kD) weight averaged molecular weight, appeared to be absorbed in similar MWDs to the administered material. Larger MWDs showed size selectivity in the absorption process. In these cases, the MWDs of absorbed material were consistently smaller than those of the administered polymers; at longer sample times as absorption proceeded, the % of larger molecules transferred to the vasculature also increased. Because mucociliary clearance from the lower airways is slow, substantial absorption of quite large molecules may be possible over 10-12 h periods following aerosol inhalation.

Although the lungs and respiratory tract can metabolize some fraction of a delivered dose, the route offers an enormous absorptive surface area capable of delivering even large compounds to the circulation at acceptable rates. Furthermore, in contrast to the situation experienced by asthma patients inhaling medicinal aerosols, drug deposition in human subjects with normal lungs can be made quite reproducible. Nevertheless, the dependence of lung metabolism and absorption upon molecular structure is poorly understood and requires further research. In the near future, it is likely that optimal formulations will be combined with modified aerosol delivery devices to maximize and achieve reproducible values for doses delivered to the lung. These will be used as alternatives to parenteral delivery of small to intermediate doses of drugs which are not absorbed via the gastro-intestinal tract.

The permeability of various human epithelial barriers to macromolecules is presently a subject of great importance to the pharmaceutical scientist, and others with interests in the optimal delivery of molecules to intra-organism targets. Inhalation and subsequent systemic absorption via the alveolar and bronchial epithelia appears to offer some advantages (*1*) over other routes. Thus, pulmonary drug delivery by aerosol is receiving increased attention because of difficulties associated with the systemic delivery of peptides, polypeptides and other macromolecules generated as products of biotechnology. The likelihood of reasonable values for their pulmonary bioavailability has led to interest in presentation of these drugs for absorption via the lung. The lung's surface area for absorption is large and in the range 35 - 140 m^2 (*2*). Also, in most societies oral inhalation is well accepted by the general population.

Early work by Wigley and co-workers (*3*) showed that a polypeptide of intermediate size, insulin (approximately 6.0 kD) (*4*) was absorbed through the lung following aerosol delivery, as evidenced by blood glucose lowering following delivery. Like much of the early work on aerosols however, it is extremely difficult to determine insulin's bioavailable fraction because the critical determinants of aerosol deposition and dosimetry in the lung (aerosol concentration and particle or droplet size distribution at the patient's oropharynx, and inhalation regime (*5*)) were not reported. Administration was accomplished with a Devilbiss No. 40 nebulizer containing an unquoted volume of 500 units/mL porcine insulin in the reservoir (there are 28.5 units anhydrous human insulin per mg of protein). Aerosol was generated and delivered to normal and diabetic humans for 5 minute time periods using unquoted operating conditions. Glucose lowering from the aerosol was roughly equivalent to that following an intravenous bolus of 0.1 units of porcine insulin per kg body weight, which is not an unrealistic dose in practice. With hindsight these results are extremely promising. The kinetics of glucose lowering, was effectively equivalent between the two modes of administration indicating the apparent rapidity of absorption from the lung.

A number of other peptide molecules are currently being explored for delivery via inhalation (*6*). Very recently, a much smaller peptide (leuprolide, about 9 amino acid residues) has been delivered by metered dose inhaler (MDI) in a characterized fashion to humans (*7*). This work revealed that about 50% of a dose deposited in the lung could be bioavailable. This value is much greater than those reported for nasal bioavailabilities of this and similar molecules (*8*). These results, and ours in the rat lung (*9*), imply that inhalation administration of some peptide and polypeptide molecules is perfectly feasible.

To enhance understanding of macromolecular absorption via the lung we have performed a series of investigations on the transepithelial transfer of fluorophore labelled polyhydroxyethylaspartamides in the rat lung. Our results will be discussed alongside some of the likely determinants of polypeptide bioavailability following delivery by inhalation.

Experimental

Poly-α,β-[N(2-hydroxyethyl)-D,L-aspartamide] containing a covalently bound fluorophore (F-PHEA) was prepared from a D,L racemic mixture of aspartic acid via polysuccinimide. Full details have been reported previously (9). Briefly, polysuccinimide was reacted, first with aminoethyl carbonyl 6-aminofluorescein, and subsequently with excess ethanolamine, to produce the ring opened, fluorophore labeled, F-PHEA (9). The product was fractionated, concentrated and desalted by gel permeation chromatography (GPC) prior to lyophilization and storage over silica gel. The F-PHEA was then delivered intracheally in solution as coarse aqueous sprays to the airways of the isolated perfused rat lung (iprl) (10). Delivery was phased with inhalation and propelled by a 25 µL bolus of drug-free propellant from a metering device (10). The circulating perfusate was sampled as a function of time after F-PHEA administration and samples were analyzed for the amount of polymer and its molecular weight distribution (MWD). Thus the airway to vasculature absorption of F-PHEA was determined as a function of time.

Characterization of F-PHEA. The content of the fluorophore was quantified by fluorimetry (λ_{ex} = 486 nm, λ_{em} = 516 nm, Luminescence Spectrometer LS 50, Perkin Elmer Corporation, Rockville, Maryland) and expressed as if it were fluorescein itself by assuming that the quantum yields of both fluorophores were identical (9). The molecular weight distribution (MWD) of the F-PHEA reported here was determined by gel permeation chromatography (GPC) using a previously calibrated, mixed-bed column (Sephacryl S-200: Sephadex G-25SF; 3:1, 12 x 300 mm, Pharmacia LKB, Uppsala, Sweden) and a mobile phase of pH 7.4 phosphate buffered saline (0.05 M phosphate; 0.15 M NaCl). The column was calibrated with three overlapping fractions of PHEA with known MWD using the curve summation method as described previously (9). The linear portion of the calibration curve relating logMW and the elution volume, V_e, covered the range 1.5 kD through 45 kD and could be expressed by the equation logMW = 6.50 - 0.19V_e. Elution of F-PHEA was monitored spectrophotometrically as the absorption at 490 nm (Model 1840 UV detector, Isco, Inc., Lincoln, NE).

Concentration and MWD of F-PHEA After Absorption. F-PHEA was determined in perfusate samples by quantitative GPC relative to a freshly prepared F-PHEA standard run on the same day. Either a mixed-bed column (12 x 300 mm Sephacryl S-200: Sephadex G-25 SF; 3:1, Pharmacia LKB) or a Separon HEMA-Bio 40 column (8 x 250 mm; 10 µm particle size, Tessek A/S, Aarhus, Denmark) was used with a 20 µL injection volume. A mobile phase of pH 7.4 phosphate buffered saline (0.05 M phosphate, 0.15 M NaCl) was supplied (Model LC-7A Bio Liquid Chromatograph, Shimadzu Corporation, Kyoto, Japan) at 0.5 or 1 mL/min. Fluorescent detection was employed (Model RF-535 Fluorescence HPLC Monitor,

Table I. Transfer to Perfusate for Several Fluorophore-labelled Molecular Weight Distributions of Poly(hydroxyethylaspartamide)[a]

Mean dose (mg)	Polymer M_w	MWD M_n	(F)	% transfer to perfusate
5.0	40,000 approx.	unknown	0.58	Undetectable
4.82	12,200	10,200	0.47	4.0
1.97	8,800	6,860	0.60	12.9
2.27	5,700	3,900	0.56	29.6
2.18	7,430[b]	4,700	1.21	11.9
0.22	4,680[b]	3,130	1.07	68.4
0.55	4,680[b]	3,130	1.07	49.0
2.16	4,680[b]	3,130	1.07	16.8
7.79	4,680[b]	3,130	1.07	10.2

Reprinted with permission from Ref. 9. Copyrighted 1990.

[a]   100 min after administration to the airways of the isolated, perfused rat lung preparation. M_w = weight mean, M_n = number mean, F = percent of fluorescein by weight, covalently bound to the polymer.

[b]   These polymers contained twice the fluorophore in order to improve analytical sensitivity.

Shimadzu; λ_{ex} = 486nm, λ_{em} = 516 nm) with a chart recorder (Fisher Recordall, Series 5000, Springfield, NJ). F-PHEA concentrations in perfusate were determined by area comparison of the eluted curves to that of the standard. Curves of logMW vs. elution volume, V_e, were constructed for each column using the integral-MWD method as described previously (9). Molecular weight distributions of absorbed material were determined from the calibration curves and the sample's chromatograph by curve summation (9).

Results

The shapes of the profiles for the amount absorbed vs. time for F-PHEA with different MWDs have been reported previously (9). In general, the amount of F-PHEA absorbed increases steadily with time although polymers with smaller MWD are absorbed more rapidly. The effect is illustrated in Table I which is reproduced from a previous paper (9). Interpretation of F-PHEA's absorption kinetics can be complicated however, either by varying the administered dose or the administered molecular weight distribution. Table I shows clearly the increasing percent absorbed of a weight averaged F-PHEA of 4.68 kD as the dose was systematically reduced. This data is evidence of the existence of carrier-mediated transfer for F-PHEA in the

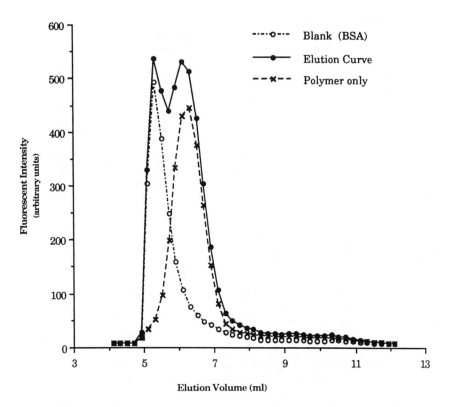

Figure 1. Typical chromatograms for (-•-)F-PHEA in the presence of serum albumin (-o-) serum albumin in perfusate and (-x-)F-PHEA following curve stripping of the data.

lung. Data collected for a further polymer was analyzed systematically for MWD at sampling times > 30 min. Samples collected at times < 30 min could not be analyzed accurately for MWD. This was not due to assay sensitivity problems with the polymer itself. The high protein content of the perfusing buffer (Krebs Hensleit with 4% Bovine serum albumin), and other biologic products in the lung perfusate, contained fluorophoric contaminants which required the use of strict controls during the assay for F-PHEA. In a wide series of experiments there was no evidence that these materials interfered with column calibrations for F-PHEA in terms of molecular weight. F-PHEA of known MWDs, for example, had identical chromatograms in the presence and absence of 4% albumin at all concentrations of interest provided fluorescence due to albumin was first subtracted from the data. Thus, protein binding did not interfere with MWD determinations. Nevertheless, for accurate F-PHEA quantification and MWD determinations, a rigorous curve stripping procedure was employed. This is illustrated in Figure 1. Control

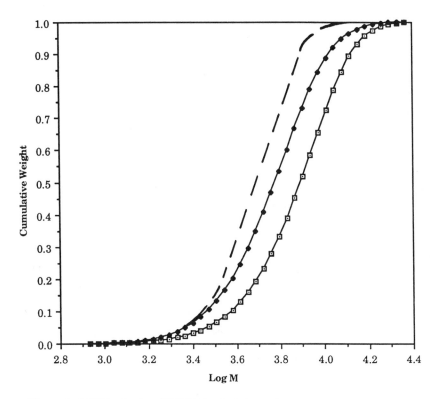

Figure 2. MWDs for (-□-)F-PHEA administered to the airways of the iprl; (-◆-) F-PHEA absorbed into the perfusate 120 minutes after administration. In these two cases the administered polymer was characterized by M_w = 8.6 and M_n = 5.3 kD. For comparative purposes the MWD is also shown for a smaller polymer which is absorbed in effectively the same MWD as it is administered: (- -)FPHEA with M_w = 4.7 kD and M_n = 3.3 kD before and after absorption.

chromatograms determined in perfusate samples containing no polymer were systematically subtracted from test chromatograms in order to eliminate the effect of matrix contaminants in the analytical procedure.

Figure 2 and Table II show results for the MWD of F-PHEA transferred to the systemic side of the iprl from the airways. These results were generated following administration of a single polydisperse batch of F-PHEA with dose held effectively constant. MWD's of absorbed F-PHEA were consistently smaller than the administered polymer. Nevertheless, these differences in MWD between administered and absorbed material (Figure 2) were, most probably, not due to metabolism during the 3-hour lifetime of the iprl (the MWD of F-PHEA recovered from the airways <u>plus that in the perfusate</u> could not be distinguished after 3 hours in the iprl from that of the administered material). The fluorophore moreover, remained bound to the polymer throughout the experiments and no fluorescent material with molecular weights < 1.0 kD was detectable following GPC of perfusate samples.

During absorption of this particular F-PHEA (M_W = 8.6 kD, M_n = 5.3 kD), through the rat lung, transfer occurred at an apparently constant rate of 110 ± 43 µg/h or $3.5 \pm 1\%$ of the administered dose per hour. Because mucociliary clearance from the lower airways occurs very slowly (1) these absorption rates convert to substantial bioavailabilities when the absorption process is extrapolated over a 12 h period [(3.5% x 12) or, around 42% may be feasible].

Discussion

In these experiments the lung appeared able to discriminate in terms of absorption rate between smaller and larger F-PHEA molecules (Tables I and II). For larger molecules, this difference in rate shows up as a continuously changing MWD in the perfusate as absorption proceeds (% <4 kD decreases with time, Table 2). However (Figure 2), this effect was not seen for a batch of F-PHEA with much smaller MWD (weight mean molecular weight, M_W = 4.7 kD; number mean molecular weight, M_n = 3.3 kD) (9). In that case, there was no difference between the elution volumes of absorbed and administered material following GPC of perfusate samples. The molecular sieving capacity of the rat lung becomes apparent for linear hydrophilic molecular weights in excess of about 7 kD. Nevertheless, molecules with sizes in excess of those reported here are absorbed at finite but slower rates by the lung (1) and a "molecular cutoff", if one exists at all, may be much larger than is presently imagined.

Delivery of drug within a fixed dosing range to the lumen of the airways can be achieved if the drug under consideration is presented as a well-defined aerosol for inhalation by a lung-normal individual (1). Under these circumstances, when the dose can be defined, factors controlling the bioavailable dose will be those, other than absorption, which represent alternative pathways for drug disappearance from the airways. In the lung-normal individual, there is reason to believe that there are fewer complications likely to interfere with the absorption process of polypeptides than there are for alternative routes of administration. In this respect, inhalation may offer fairly reproducible absorption kinetics and thus have some advantages. For example, obvious differences between the lung and the gastrointestinal tract include the lack of dietary complications in the lung, the low levels of extracellular enzymes and the absence of quantitative and qualitative interindividual metabolic differences due to the presence of different flora and fauna in the respiratory tract.

The key issues affecting pulmonary bioavailability are shown in Scheme I. Both the conducting airways and the alveoli are likely to show significant rates of drug absorption. The greater importance of the alveolar regions stems mainly from the fact that mucociliary clearance is considerably slower from the lung periphery than it is from the upper airways (13). For larger molecules, the increased duration of residence in the lower lung should lead to a greater opportunity for absorption from the alveoli. In Scheme I, mucociliary clearance to the gastrointestinal tract,

GIT, is considered to be lost from the absorption site, as is metabolic clearance and sequestration in various cell types and membranes (*12, 14*). It is clear from Scheme I that the relative rates of the various processes will define the bioavailable fraction of the dose and understanding those factors which control pulmonary absorption kinetics is obviously the key to enhancing bioavailability via the lung. In a recent book (*15*) the molecular dependence of lung binding and metabolism was considered alongside the parallel processes of absorption, clearance and dissolution in the lung (*14*). Some key features of this work will be repeated as it relates to the systemic delivery of polypeptides.

The data presented in Table III is restricted to that collected using the in situ rat lung technique by Schanker et al. (*16-23*). The molecular weight dependence is shown without regard for slight differences in solute lipophilicity and extended to consider the absorption of a limited number of macromolecular solutes. However, the Table is also restricted to hydrophilic solutes, like proteins and polypeptides, which are likely to pass either via tight junctions in the epithelial cell layer or by endocytosis (*14*). While membrane permeabilities are clearly variable, the absorption half-lives for insulin and dextran (20 kD) are consistent with the values seen here for PHEA (20 kD dextran has an absorption half-life of about 12 h). Other work on the pulmonary absorption of macromolecules is sparse. Table I however, summarizes the net transfer or absorption of different polydisperse PHEA molecular weight distributions in the isolated lung. This data has been published in detail elsewhere (*9*) but was also included here to show the likely dependence of absorption kinetics upon macromolecular MWD. Absorption was characterized with time and, also (Table I) 100 min after dosing to the airways of the isolated perfused rat lung as aqueous sprays (*10*). Interestingly, one of the smaller molecules shows a dose dependence rather like that reported for disodium cromoglycate or phenol red in the in situ rat lung (*24, 25*). These degrees of dose dependence, when taken alongside the fact that doses administered by inhalation are often small, may mean that enhanced bioavailability of small metered doses could well result from the presence of, as yet, ill-defined carrier systems in the lung.

While all the MWDs of PHEA (Table I) delivered to the isolated rat lung have been absorbed to some degree, there is a sieving effect which differentiates between small and large molecules in terms of the rates at which they are absorbed. PHEA is minimally metabolized by the isolated lung. This is believed to be due to its mixed alpha and beta "peptide bond-like" linkages and R and S stereochemistry in the aspartate moieties (*9*). This rather resistant structure is unlikely to be typical of biologically active polypeptides but is ideal for the purpose of determining the structural dependence of the pulmonary absorption process. Some peptide molecules like leu- and met-enkephalin are rapidly metabolized in lung by aminopeptidases which are probably membrane bound at the endothelium (*26, 27*). However, studies are usually performed to discern lung metabolism either in cell homogenates or following delivery direct to the pulmonary blood supply and endothelia. Enzyme systems and their locations in the lung have not been studied widely, nor are there many appropriate models other than isolated or in situ lung systems for screening metabolism during absorption. The fact that insulin (*3*) and

**Table II. The Mean % of the Absorbed F-PHEA with Molecular
Weight < 4.0 kD[a]**

SAMPLE TIME (min)	MEAN% <4000 Daltons
30	29.66
60	27.61
100	26.09
150	26.50
180	24.97

[a] Determined by GPC of samples withdrawn from the circulating perfusate of the isolated rat lung preparations; administered polymer was characterized by M_w = 8600, M_n = 5300 Daltons.

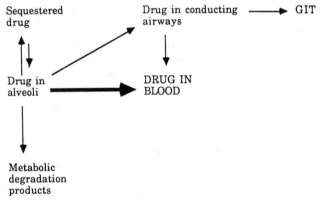

Scheme I

**Table III. Absorption Half-lives (Minutes)
of Some Lipid Insoluble Compounds
Alongside Some Hydrophilic Macromolecules [a]**

Solute	Molecular Weight	Half-life
Procainamide ethobromide (PAEB)	264	70.0
N-Acetyl PAEB	306	38.5
Benzylpenicillin	334	33.0
Sucrose	342	87.0
Tc-Diethylenetriaminepentaacetate	492	30.9
Cyanocobalamin	1,355	180.0
Inulin	5,250	225.0
Dextran	20,000	688.0
	75,000	1670.0

[a] Absorption data from ref. 16-23.

leuprolide (*7*) have substantial bioavailabilities following inhalation may indicate that whole organ experiments must be performed in order to determine realistic values for first-pass metabolism in the lung.

CONCLUSIONS

Our work, and that of others, has established that macromolecules with sizes up to about 20 kD may feasibly be administered systemically via inhalation. The lung can be reached almost instantaneously by aerosol and, due to the slow clearance of aerosols from the lower lung, even compounds with very small absorption rates can probably be transferred to the systemic circulation in significant quantities over 10-12 hour periods. Presently unknown however, is the precise nature of the absorption processes for large molecules and its dependence upon molecular structure. In the future, we must also determine the kinetics of metabolism for peptide and protein molecules with useful biological activity. Nevertheless, intelligent use of this route of administration should enable pharmaceutical scientists to avoid many of the first pass metabolism problems inherent with alternate routes. Clearance from the site of absorption is more rapid in the nose than it is in the lung and if the doses required for systemic activity are of the order of milligrams or less, present dosage forms and minor improvements in existing technology are quite capable of delivering compounds to the lung prior to their absorption. In the near future, it is likely that optimal formulations will be combined with modified aerosol delivery devices to maximize and achieve reproducible values for dosage to the lung. These will be used as alternatives to parenteral delivery for small through intermediate dose drugs.

Acknowledgments

We are grateful to Eli Lilly and Company for their support of this work.

Literature Cited

1. Byron, P.R. *Adv. Drug Deliv. Rev.* **1990**; *5*, 107.
2. Hollinger, M.A. *Respiratory Pharmacology and Toxicology,* W.B.Saunders Company, Philadelphia, PA, 1985, pp. 1-20.
3. Wigley, F.M.; Londono, J.H.; Wood, S.H.; Shipp, J.C.; Waldman,R.H. *Diabetes* **1971**; *20,* 552.
4. *Merck Index 9th Ed.,* Merck, Rahway, NJ, 1976; pp. 659.
5. Byron,P.R. In *Respiratory Drug Delivery;* Byron, P.R., Ed.; CRC Press, Boca Raton, FL, 1990; pp. 143.
6. Ahmed, T. In *Respiratory Drug Delivery;* Byron, P.R., Ed.; CRC Press, Boca Raton, FL; 1990; pp. 208.
7. Adjei, A.; Garren, J. *Pharm. Res.* **1990**; *7,* 565.
8. Chan, R.L.; Henzl, M.R.; LePage, M.E.; LaFargue, J.; Nerenberg,C.A.; Anik, S.; Chaplin, M.D. *Clin. Pharmacol. Ther.* **1988**; *44,* 275.
9. Niven, R.W.; Rypacek, F.; Byron, P.R. *Pharm Res.* **1990**; *7,* 990.

10. Byron, P.R.; Niven, R.W. *J. Pharm. Sci.* **1988**; *77,* 693.
11. Gonda, I.; Byron, P.R. *Drug Dev. Ind. Pharm.* **1978**; *4,* 243.
12. Crooks, P.A.; Damani, L.A. In *Respiratory Drug Delivery;* Byron,P.R., Ed; CRC Press, Boca Raton, FL, 1990; pp. 62.
13. Byron, P.R. *J. Pharm. Sci.* **1987**; *75,* 433 and 1207.
14. Byron, P.R.; Phillips, E.M. In *Respiratory Drug Delivery;* Byron, P.R., Ed; CRC Press, Boca Raton, FL, pp. 108.
15. Byron, P.R. (Ed.) *Respiratory Drug Delivery,* CRC Press, Boca Raton, FL., 1990.
16. Lanman, R.C.; Gillilan, R.M.; Schanker, L.S. *J. Pharmacol. Exp.Ther.* **1973**; *187,* 105.
17. Enna, S.J.; Schanker, L.S. *Am. J. Physiol.* **1972**; *223,* 1227.
18. Burton, J.A.; Schanker, L.S. *Proc. Soc. Exp. Biol. Med.* **1974**; *145,* 752.
19. Burton, J.A.; Schanker, L.S. *Xenobiotica* **1974**; *4,* 291.
20. Burton, J.A.; Gardiner, T.H.; Schanker, L.S. *Arch. Environ. Health* **1974**; *29,* 31.
21. Schanker, L.S.; Burton, J.A. *Proc. Soc. Exp. Biol. Med.* **1976**; *152,* 377.
22. Schanker, L.S.; Hemberger, J.A. *Biochem. Pharmacol.* **1983**; *32,* 2599.
23. Brown, R.A.; Schanker, L.A. *Drug Metab. Disp.* **1983**; *11,* 392.
24. Enna, S.J.; Schanker, L.S. *Life Sci.* **1973**; *12,* 231.
25. Gardiner, T.H.; Schanker, L.S. *Xenobiotica* **1974**; *4,* 725.
26. Gillespie, M.N.; Krechniak, J.W.; Crooks, P.A.; Altiere, R.J.; Olson, J.W. *J. Pharmacol. Exp. Ther.* **1985**; *232,* 675.
27. Llorens, C.; Schwartz, J.C. *Eur. J. Pharmacol.* **1981**; *69,* 113.

RECEIVED May 6, 1991

POLYMERIC DRUG DELIVERY

Polymeric Drug Delivery

One of the most common methods for using polymers to deliver drugs is to physically blend or mix the drug with the polymer and have it release by diffusion through the polymer, dissolution from the inert polymer, or degradation or erosion of the polymer. In these delivery systems, the polymer is neither biologically active nor chemically attached to the drug. Most of the polymeric drug delivery systems commercially available and being developed utilize polymers simply as the matrix for delivering the drug. These include oral, parenteral and transdermal delivery systems, and the polymers encompass both biodegradable and nonbiodegradable materials. The papers in this section represent some of the research being devoted to new biodegradable polymers for drug delivery and their mechanism for drug release, polymers with both hydrophilic and hydrophobic properties, and hydrogel polymers for oral drug delivery.

The first paper in this section (Chapter 14) describes the synthesis and characterization of biodegradable poly(phosphoester-urethanes) which are so versatile that they can be used either as a polymer-drug conjugate or a matrix for delivery of antitumor agents. In Chapter 15, a new series of biodegradable polymers classified as "pseudopeptides" is presented. The synthesis and thorough characterization of these series of polymers is discussed. The potential for these materials to be useful both as biomaterials and drug delivery matrices is indicated. The next paper in this section (Chapter 16) presents a mathematical model for the erosion characteristics of poly(ortho ester)s which have been extensive evaluated for drug delivery because of their bioerodible properties. The correlation of this model with experimentally obtained polymer molecular weight changes and drug release is quite accurate. In Chapter 17 is shown the results of blending together two well known materials in drug delivery: the biodegradable poly(D,L-latic acid) and the nonbiodegradable ethylene/vinyl acetate copolymer. The degradation and pore-forming characteristics may prove useful in some unique type of drug delivery.

The next chapters deal with nonbiodegradable polymers. In Chapters 18 and 19, the synthesis and charactization of polymers containing amphiphilic networks is described. These polymer which have both hydrophilic and hydrophobic polymer segments may afford some interesting properties for drug delivery. Chapter 20 presents the results obtained with a crosslinked polyacid which was evaluated for oral delivery of insulin. The swelling of this hydrogel material in the intestine rather than the stomach was to protect the drug from the hostile gastric conditions and release it where it could be absorbed. The last chapter in this section (Chapter 21) describes another hydrogel material but this time the polymer is designed to be retained in the stomach for drug delivery. The preparation of the polymer and the evaluation of its retention in the stomach are discussed.

Chapter 14

Synthesis and Characterization of Hydrolytically Labile Poly(phosphoester—urethanes)

F. Y. Shi[1], L. F. Wang, E. Tashev[2], and K. W. Leong[3]

Department of Biomedical Engineering, Johns Hopkins University, Baltimore, MD 21218

Biodegradable poly(phosphoester-urethanes) containing bisglycophosphite as the chain extender were synthesized. Methylene bis-4-phenyl isocyanate (MDI) and toluene diisocyanate (TDI) were initially used as diisocyanates. Since there was a concern that the degradation products could be toxic, the ethyl 2,6-diisocyanatohexanoate (LDI) was synthesized and replaced the MDI (or TDI). The hydrolytic stability and solubility of these polymers were tested. Preliminary release studies of 5-fluorouracil from MDI based poly(phosphoester-urethane) and methotrexate from LDI based poly(phosphoester-urethane) are also reported.

Polyurethane has been extensively studied and widely used as a biomedical material since the early 1960's (1). Recent advances of this class of materials has been reviewed by Gogolewski (2) and Cooper (3). It is a fascinating class of polymers which through the variations of the soft and hard segments and the chain extenders can exhibit diverse physicochemical properties. The polymers have good tissue and blood biocompatibility and they are stable in the body. With strong mechanical strength, including elastomeric properties, they are prime candidates for medical applications where inertness and stability are crucial.

Our interest in the past few years has been on biodegradable polymers. We have been evaluating the potential of poly(phosphoesters) as degradable biomaterials (4).We were attracted to this class of polymers because the phosphoester bond in the backbone is cleavable under physiological conditions, the presence of the P-O-C group would facilitate fabrication, and the versatile chemical structure affords a wide

[1]Current address: Institute of Polymer Chemistry, Nankai University, People's Republic of China
[2]Current address: Central Laboratory for Polymers, Bulgarian Academy of Sciences, Bulgaria
[3]Corresponding author.

range of physical properties. In addition, the pentavalency of phosphorus furnishes a functional side chain. Initially suggested by Penczek, these polymers allow direct linkage of bioactive compounds to the phosphorus atom to form a pendant controlled release system (5). To combine biodegradability and desirable mechanical strength, we examined the synthesis and characterization of polyurethanes which include a phosphoester bond in the backbone.

We introduced the biodegradability into the polyurethane by incorporating a bisglycolphosphite (BEG) as part of the chain extender. To demonstrate the degradability of these polymers, we initially chose the toluene diisocyanate (TDI) or methylene diphenyl isocyanate (MDI) as the counterpart since they are the two most commonly used diisocyanates in polyurethane chemistry. With concerns that these aromatic diisocyanate structures might yield toxic products when degraded, efforts were then made to synthesize a diisocyanate which after polyurethane formation would breakdown into nontoxic residues. To achieve this, we started with the conversion of the amino acid lysine into a diisocyanate. It is envisioned that polyurethanes synthesized from this diisocyanate will regenerate lysine when biodegraded, evading any toxicity problems usually associated with aromatic diamines.

Chemical structure of monomers and intermediates was confirmed by FT-IR and FT-NMR. Molecular weight distribution of polymers was assessed by GPC and intrinsic viscosity. The thermal property was examined by differential scanning calorimetry. The hydrolytic stability of the polymers was studied under in vitro conditions. With controlled drug delivery as one of the biomedical applications in mind, release studies of 5-fluorouracil and methotrexate from two of these polymers were also conducted.

Experimental

The different components for the polyurethane syntheses are shown in Table I.

Materials and Purification. Chemicals were purchased from Aldrich chemical company and used as received unless otherwise noted: 1,1,1,3,3,3-hexamethyl disilazane, ethylene glycol, triphosgene, poly(ethylene oxide) (MW = 600), poly(tetramethylene oxide) (MW = 1000), poly(caprolactonediol) (MW = 530), toluene diisocyanate (TDI), anhydrous ethanol (Barker Analyzed), L-lysine monohydride (Sigma) and methylene bis-4-phenyl isocyanate (MDI) (Kodak). Ethyl ether (Barker Analyzer), triethylamine and dimethyl acetamide were respectively dried with sodium, calcium hydride and barium oxide overnight, and then distilled. Thionyl chloride and diethylphosphite were distilled before use.

Synthesis. Bis(2-hydroxyethyl)phosphite was synthesized by adopting the procedures described by Borisov and Troev (6). Briefly, a molar ratio of 4:1 of ethylene glycol (73 mL; 1.3 mole) to diethylphosphite (43 mL; 0.33 mole) was placed in a round bottom flask equipped with a reflux condenser and a thermometer. A two mL fresh solution of sodium methoxide was added dropwise through a dropping

Table I. The chemical structures used in the synthesis of poly(phosphoester-urethanes)

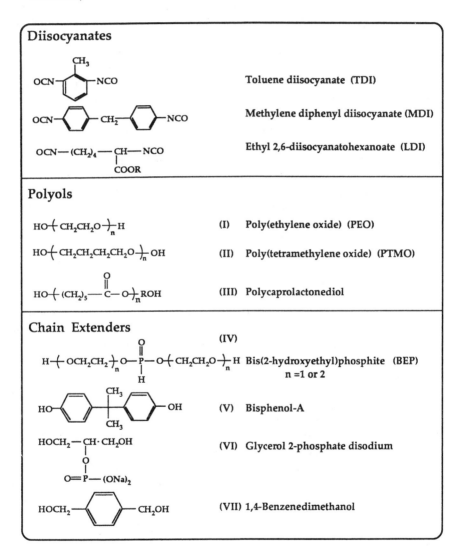

funnel over a period of 2 minutes. The reaction mixture was kept at 135-140°C until the theoretical amount of ethanol was removed. The temperature was then raised to 152°C while a vacuum of 20 mm Hg was applied for 3 hours to remove the unreacted ethylene glycol and impurities. The product was a colorless viscous liquid.

To obtain the ethyl 2,6-diisocyanatohexanoate (LDI), the strategy described by Katsarava et al (7) was adapted (Scheme 1). L-lysine was converted to lysine ethyl ester by refluxing a mixture of 50 g of L-lysine monohydrochloride, 20 mL of thionyl chloride, and 300 mL of anhydrous ethanol. When the reaction mixture turned clear, after approximately 4 hours of reflux, the excess ethanol was distilled off to yield the solid product L-lysine ethyl ester. The L-lysine ester was then recrystallized in ethanol. For the disilazane intermediate, 25 g of L-lysine ethyl ester and 200 mL of 1,1,1,3,3,3-hexamethyldisilazane were placed in a round bottom flask with a condenser protected by a drying tube. While agitated, the suspension was maintained at 120°C for 24 hours or more until the mixture turned clear. The excess 1,1,1,3,3,3-hexamethyldisilazane was removed and the crude product purified by vacuum distillation at 100°C and 0.1 mmHg to yield a colorless liquid. The final conversion to the diisocyanate was conducted in an Atmosbag purged with dry nitrogen. Great precaution was taken to exclude moisture from the system. Twenty five mL of the purified bis(trimethylsilyl)L-lysine was dissolved in 200 mL of anhydrous ethyl ether in the presence of 19.4 mL of triethylamine. While the mixture was cooled at -20°C, 15 g of triphosgene dissolved in 100 mL of anhydrous ethyl ether was added dropwise over a period of 30 minutes. The mixture was warmed to room temperature in 2 hours and allowed to react for 40 hours. After the triethylamine hydrochloride was removed by filtration, the solution was evaporated to dryness. The residue was then purified by distillation at 120°C and 0.1 mmHg. The typical yield of ethyl 2,6-diisocyanatohexanoate from L-lysine monohydrochloride was 50%.

To obtain the polyurethanes, typically a prepolymer was first prepared by reacting the diisocyanate with various diols in dimethylformamide or dimethylacetamide in a two to one molar ratio at 100-110°C for two hours under nitrogen atmosphere. A solution of chain extenders, such as BEP, was then added to the prepolymer reaction mixture and further reacted another three hours. The polymer was isolated by quenching the reaction mixture in cold water. Fine white powder was obtained with a typical yield of around 90%.

To explore the feasibility of these polymers as a pendant delivery drug delivery system, 5-fluorouracil activated by 1,1,1,3,3,3-hexamethyl disilazane was coupled to the phosphite via the Todd reaction as shown in Scheme 2. Briefly, the polymer (PPU-7) solution in dimethylacetamide was mixed with a 1:1 v/v solution of carbon tetrachloride and methylene chloride containing 0.5 volume percent of triethylamine (TEA) as catalyst. The activated drug dissolved in methylene chloride was added slowly into the mixture and allowed to react for 48 hours at room temperature. The product was isolated by precipitation into carbon tetrachloride.

Characterization. The FT-IR spectra were obtained on a Perkin Elmer 1600 Series machine by coating the samples on a NaCl plate. The purity of BEP was examined by FT-NMR at 400 Mhz in deuterated chloroform on a Varian XL 400. The

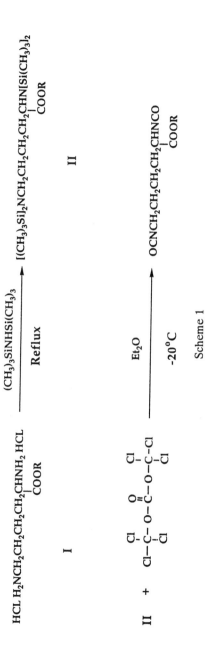

Scheme 1

molecular weight properties of the polymers were characterized by gel permeation chromatography and intrinsic viscosity. The GPC was performed in a Hewlett-Packard 1090M system equipped with a diode array and a refractive index refractometer. Samples were eluted through three PL-Gel columns in series at 40°C. Polystyrene standards were used to calibrate the system. The intrinsic viscosity measurements were done in Fenske viscometers in DMF at 30°C. The thermal properties of the polymers were analyzed by DSC on a Mettler TA3000 system at a heating rate of 10°C/minute under nitrogen after the thermal history of the samples was erased in preliminary heating cycles. To estimate the hydrolytic stability of the polymers, hydrolysis experiments were conducted in 0.1M pH 7.4 phosphate buffer at 37°C as well as in 0.1M NaOH at room temperature as an accelerated test. The MDI based polyurethanes were tested as discs, with a dimension of 1 cm in diameter and 1 mm thick, fabricated by compression molding at room temperature and at 75 Kpsi. The LDI based polyurethanes were tested in the form of a thin film between 20 to 50 microns thick, obtained by solvent casting from DMA. The degradation kinetics was followed by UV spectrophotometry and corroborated by mass loss measurements.

In vitro release study. For 5-fluorouracil which had been covalently coupled to MDI based polyurethane, the release study was done on just the polymer-drug particles. Samples were placed in Biorad chromatography columns (1.5 x 10 cm) containing 10 mL of 0.1M pH 7.4 phosphate buffer. The buffer solutions were incubated at 37°C and changed periodically for UV spectroscopic analysis. Release of methotrexate from LDI based polyurethane was done on a thin film obtained by solvent casting from DMA. A 10 weight percent drug loading level was used. After removal of solvent in a vacuum oven, samples were punched out from the sheet by a corkborer and placed in phosphate buffer for release study as described above.

RESULTS

TDI/MDI Based Poly(phosphoester-urethanes)(PPU). The structures of the different TDI or MDI based poly(phosphoester-urethanes) synthesized are shown in Table II. The IR spectra confirmed the presence of urethane bonds in all the samples. The polymers dissolved well in polar solvents such as DMF, DMA, and DMSO; several of them also dissolved in chloroform and methylene chloride (Table III). It is known that the P-O-C group has a plasticizing effect and enhances solubility in common organic solvents (8). Hence it appeared that the inclusion of the bisglycol phosphite in the urethane backbone did improve its solubility in organic solvents. The GPC analysis indicated a high molecular weight for these polymers, as high as several millions according to the polystyrene standards. Although the GPC values are only relative, the high intrinsic viscosities seemed to support the high molecular weight nature of these polymers. The molecular weight distribution of these polymers was highly dispersed. Shown in Figure 1 is a typical dispersion of the polymers. The polydispersity, sometimes as high as 30, may be due to the presence of impurities in the BEP, since the transesterification of diethylphosphite and ethylene glycol was likely to produce a number of side products (6). The NMR of BEP

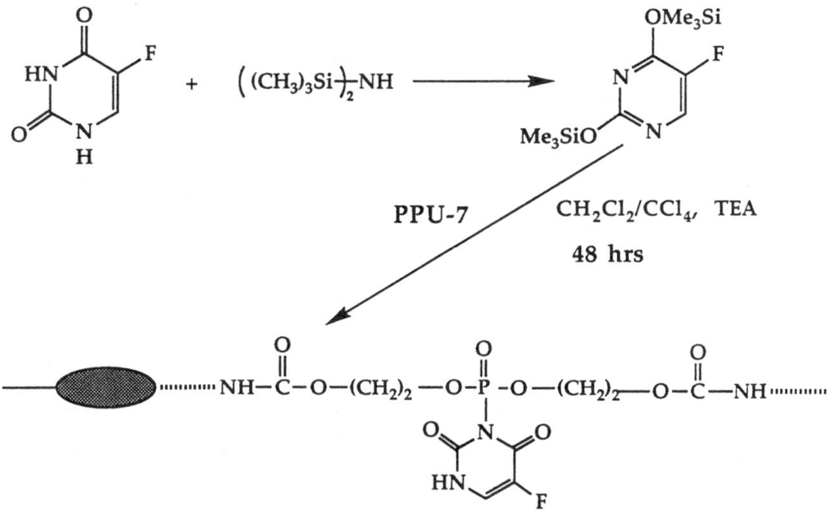

Scheme 2

Table II. Composition and intrinsic viscosity of TDI- or MDI-based polyurethanes

Polymer	TDI or MDI* (molar ratio)	DIOL (molar ratio)				$[\eta]^{30 C}_{CHCl_3}$ (dl/g)
		IV	V	I	II	
PPU-1	2	2				-
PPU-2	2	1		1		1.10
PPU-3	2	1	1			0.92
PPU-4	2	1	0.5	0.5		0.50
PPU-5	2		1	1		1.16
PPU-6	2	1			1	1.95
PPU-7	2*	1.5		0.5		-
PPU-8	2*	2				-

Table III. Solubility of MDI- or TDI-based poly(phosphoester-urethanes)

Polymer	DMF	DMSO	CH₃Cl	CH₂Cl₂	Dioxane
PPU-1	-	+	-	-	-
PPU-2	+	+	-	-	-
PPU-3	+	+	-	-	-
PPU-4	+	+	-	-	-
PPU-5	+	+	-	-	-
PPU-6	+	+	+/-	-	-
PPU-7	+	-	swell	+	swell
PPU-8	+	-	swell	+	swell

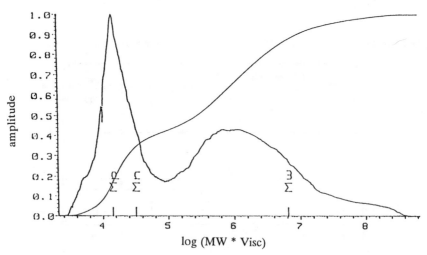

Figure 1: Molecular weight distribution of PLU-1 poly(phosphoester-urethane).

showed two sets of phosphite hydrogen around 6.0 and 7.9 ppm. The presence of P-OH is also suggested by a peak at 10.3 ppm. The purification of BEP by vacuum distillation has proved to be difficult as the compound has a high boiling point.

The in vitro degradation profiles of several TDI poly(phosphoester-urethanes) are shown in Figure 2. It is not possible from this study to correlate the decomposition kinetics with the chemical structure, except for the fact that biodegradability is demonstrated. The in vitro release of 5-FU from PPU-7 is shown in Figure 3. After an initial burst, a reasonably steady and sustained release followed. The UV spectrum of the released 5-FU was identical to that of pure 5-FU, suggesting the chemical integrity of the drug.

LDI Based Poly(phosphoester-urethanes)(PLU). Because it is suspected that the biodegradable TDI or MDI based poly(phosphoester-urethanes) might be toxic, subsequent efforts concentrated on synthesizing polyurethanes which might break down into inert substances. The IR spectra of the diisocyanate LDI monomer, and its bis(trimethylsilyl) intermediate are shown in Figure 4. The characteristic peaks of the intermediate, Si-C stretching vibration and CH_3 deformation vibration of $(CH_3)_3Si$-, were observed at 842 and 1250 cm^{-1}, respectively. The chemical structure of the LDI monomer was confirmed by the intense absorbance at 2261 cm^{-1} attributed to N=C=O stretching vibration. After polymerization, the isocyanate peak disappeared, replaced by urethane peaks at 1723 and 3368 cm^{-1} due to the C=O and N-H stretching vibrations. A key step to the success of the polymerization was the isolation and purification of the bis(trimethylsilyl) intermediate. We found that without purifying that intermediate, the polymer molecular weight and yield were poor.

The composition of several poly(phosphoester-urethanes) based on this lysine diisocyanate is shown in Table IV. They have estimated high molecular weights, good film-forming properties, and elastomeric characteristics. The DSC showed a melting transition around 120-125°C. Although polyurethanes based on curing of prepolymer poly(L-lactide-co-caprolactone) with LDI exhibit glass transition temperatures around 2 to 8°C (9), no Tg's were observable above -60°C for these polymers. The reason might be the presence of moisture absorbed in the course of DSC studies, which would mask the appearance of the Tg. In comparison with polyurethanes based on aromatic diisocyanates and diamine (or diol) extenders, our phosphite extender is susceptible to stronger hydrogen bonding with water molecules. These water molecules might not have been completely removed even in the second DSC scan.

The in vitro mass loss of LDI based poly(phosphoester-urethane) films is shown in Figure 5. There was a high degree of scattering in the daily mass loss data, and hence only the cumulative value at 7 days is shown. Since each time point requires a different sample, it is not clear if this scattering suggests that the polymer is heterogeneous macroscopically. The general trend of the biodegradation rate however correlates well with the chemical structure. PLU-4 contains no degradable linkage in the backbone and is expected to be stable. The modest weight loss is possibly caused by the cleavage of the disodium phosphate in the side chain. PLU-5 showed a mass loss because of the presence of the biodegradable BEP. PLU-2

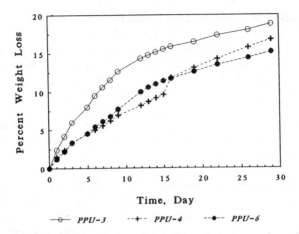

Figure 2: Hydrolytic degradation of poly(phosphoester-urethanes) based on TDI.

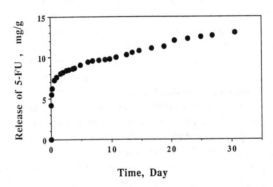

Figure 3: In vitro pendant release of 5-FU from PPU-7.

Table IV. Composition and properties of LDI-based poly(phosphoester-urethanes)

Polymer	LDI (molar ratio)	DIOL (molar ratio)						Mw 10^5	Tm, $^\circ$C
		I	II	III	IV	VI	VII		
PLU-1	2	1			0.5		0.5	63.8	124
PLU-2	2			1	0.5		0.5	73.7	-
PLU-4	2		1			0.5	0.5	2.72	125
PLU-5	2		1		0.5		0.5	65.6	120

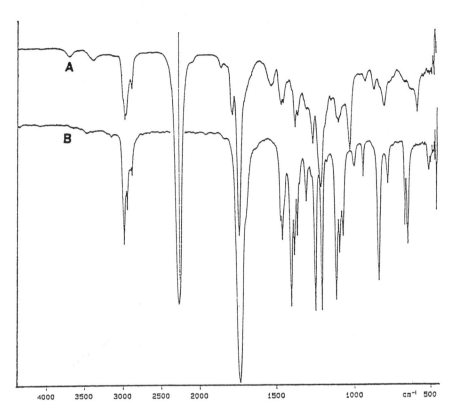

Figure 4: IR spectra of the diisocyanate LDI monomer(A) and the bis(trimethylsilyl) intermediate(B).

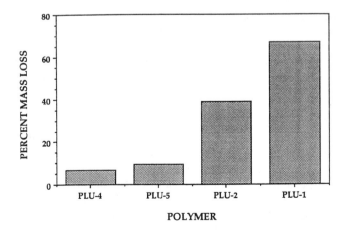

Figure 5: Weight loss of lysine-based poly(phosphoester-urethanes) at 7 days in 0.1M pH 7.4 phosphate buffer at 37°C.

showed a higher mass loss than that of PLU-5 because of the substitution of PTMO with polycaprolactone, which is degradable in itself. Finally, when the PTMO was replaced by the more hydrophilic PEO, an even higher mass loss was observed in PLU-1. In addition to the mass loss data, GPC was performed to verify that chain cleavage occurred. Shown in Figure 6 is an overlay of the molecular weight distribution of PLU-2 before and after degradation.

In vitro release of methotrexate from the LDI based films is shown in Figure 7. The release was fast since the drug was imbedded in a thin film, in the order of 50 to 100 microns. The release profile was typical of one seen in diffusion controlled systems. The scattering of the release data was again reminiscent of the mass loss results, and the cause of that is not certain at this point.

Discussion

Biodegradable polyurethanes have been proposed and studied before (*9-12*). The difference in our study is the inclusion of a phosphoester linkage instead of the commonly used polyester component. This seems to provide more flexibility as the side chain of the phosphate or phosphonate can be varied. For controlled drug delivery applications, drugs can be linked to this site to form a pendant delivery system. Moreover, for certain medical applications, fast degradation rate is obtainable by the introduction of these hydrolyzable phosphoester bonds. With the LDI based polyurethanes, drugs or other compounds of interest can also be coupled to the ester side chain of the lysine portion.

While the TDI or MDI based polyurethanes are satisfactory in biomedical applications because they are biostable, the situation might be completely different if they are made biodegradable. Suspecting the safety of the breakdown products of such polyurethanes, we did some preliminary toxicity studies in a separate study (*13*). We examined the effects of one biodegradable TDI based polyurethane in the form of microparticles on mouse macrophages. Since particulates are often generated in the biodegradation of an implant and internalized by macrophages, we thought this would be a meaningful toxicology test. It was found that the rate of DNA synthesis, which reflects the proliferative capacity, and the total protein synthesis, which indicates the metabolic activity, of the macrophages, were impaired when a sample (polyurethane particles two microns or smaller) concentration of 0.5 mg/mL was tested. Poly(L-lactide)(PLA) particles tested as controls had no effects on the functional activity of the macrophages.

Urethane hydrolyzes into an amine, an alcohol, and carbon dioxide. So the possible degradation products of a poly(phosphoester-urethane) are diamines, diols, phosphates, carbon dioxide, and even ureas. Urea is possible because the isocyanate is extremely sensitive to moisture, which would convert the isocyanate to an amino group. One is therefore bound to have traces of diamine in the polymerization that leads to a urea bond in the backbone. We think the cytotoxicity seen in the macrophage functional assay comes from the TDI structure.

Given the versatility of polyurethanes, for instance by manipulating the nature of the soft segment or the chain extender, the introduction of a hydrolytically

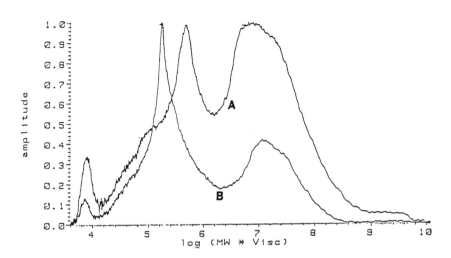

Figure 6: Molecular weight distribution of PLU-2 before (A) and after (B) degradation in 0.1M pH 7.4 phosphate buffer at 37°C.

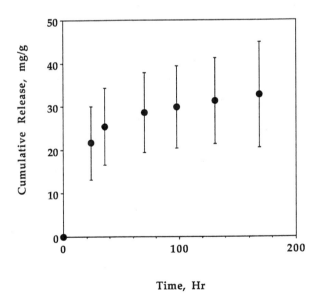

Time, Hr

Figure 7: Release of methotrexate from PLU-2 thin film in 0.1M pH 7.4 phosphate buffer at 37°C.

cleavable bond into the backbone should conceive an interesting class of biodegradable polymers. If the LDI based poly(phosphoester-urethanes) can pass the biocompatibility scrutiny, they will be fascinating biomaterials.

Acknowledgment

Support to this research is provided by the Whitaker Foundation and NIH--EY07701.

References

1. Lyman, D.J. *Rev. Macro. Chem.* **1966**, *1*, 355.
2. Gogolewski, S. *Colloid Polym. Sci.* **1989**, *267*, 757.
3. Lelah, M.D.; Cooper, S.L. *Polyurethanes in Medicine;* CRC press: Boca Raton, FL, 1986.
4. Richards, M.; Dahiyat, B.; Arm, D.; Brown, P.; Leong, K. *J. Biomed. Mat. Res.* (submitted).
5. Penczek, S.; Lapienis, G.; Klosinski, P. *Pure & Applied Chem.* **1984**, *56*, 1309.
6. Borisov, G.; Troev. K. *European Polym. J.* **1973**, *9*, 1077.
7. Katsarava, R.D.; Kartvelishvili, T.M. *Dokl. Akad. Nauk SSSR* **1985**, *281(3)*, 591.
8. Sandler, S.R.; Karo, W. *Polymer Synthesis;* Academic press, New York, NY, 1974; Vol.1, 367-445.
9. Bruin, P.; Veenstra, G.J.; Nijenhuis, A.J.; Pennings, A.J. *Makromol. Chem., Rapid Commun.,* **1988**, *9*, 589.
10. Gogolewski, S.; Pennings, A.J. *Makromol. Chem., Rapid Commun.* **1982**, *3*, 839.
11. Smith, R.; Williams, D.F.; Oliver, C. *J. Biomed. Mater. Res.* **1987**, *21*, 1149.
12. Bruin, P.; Smedinga, J.; Pennings, A.J.; Jonkman, M.F. *Biomaterials,* **1990**, *11*, 291.
13. Tanner, K.; Kadiyala, S.; Leong, K.; Frondoza, C. *Annual Meeting of Society for Biomaterials,* **1991**.

RECEIVED May 13, 1991

Chapter 15

Desaminotyrosyl–Tyrosine Alkyl Esters
New Diphenolic Monomers for the Design of Tyrosine-Derived Pseudopoly(amino acids)

Joachim Kohn

Department of Chemistry, Rutgers University, New Brunswick, NJ 08903

In peptide chemistry, peptide analogs containing nonnatural backbone linkages are classified as "pseudopeptides". In analogy, we used the term "pseudopoly(amino acid)" to denote polymers in which amino acids are linked together by bonds other than conventional peptide bonds. Here we summarize our studies on the identification of desaminotyrosyl-tyrosine alkyl esters as new diphenolic monomers for the design of amino acid derived polyiminocarbonates and polycarbonates. In particular, desaminotyrosyl-tyrosine hexyl ester was found to lead to mechanically strong polymers with good engineering properties. These new pseudopoly(amino acids) appear to be promising materials for a number of biomedical applications.

It is a well established observation that the mechanical strength and toughness of a polymer can be increased by the incorporation of aromatic monomers into the polymer backbone. For example, some of the strongest polymers known today (Kevlar and PEEK) are both exclusively derived from aromatic monomers. Yet, among the biomaterials, aromatic monomers are very rarely used and high molecular weight poly(L-lactic acid) (a completely aliphatic polymer) is the strongest degradable implant material currently available (1).

Since extremely strong, yet degradable polymers would present an important addition to the range of currently available polymeric biomaterials, we attempted to identify aromatic monomers that could be safely used in the design of high strength biomaterials. Unfortunately, the limited biocompatibility of many industrially used aromatic monomers imposes severe restrictions on the use of aromatic backbone structures in degradable polymers intended for medical applications. Diphenols, for instance, which are necessary intermediates in polyurethane, polycarbonate, and polyester chemistry are often too toxic to be considered in the design of a degradable

0097–6156/91/0469–0155$06.00/0

biomaterial. In a detailed comparison of relevant toxicological data, we identified Bisphenol A (BPA) as probably one of the least toxic monomers among the industrial diphenols (2). Due to the use of BPA in the fabrication of polycarbonate based food containers, baby bottles and kitchen utensils, the toxicological properties of BPA have been carefully studied (3): BPA has a low level of oral toxicity and was found to be noncarcinogenic in a National Cancer Institute bioassay (4). Most significantly, BPA-derived polyiminocarbonates produced only a relatively mild "foreign body reaction" when implanted subcutaneously in mice, rabbits and rats (2,5). Similar results were recently observed for BPA-containing polyphosphoesters that also showed good tissue compatibility in preliminary implantation studies (6). On the other hand, we found BPA to be highly cytotoxic in vitro using a chick embryo fibroblast tissue culture assay (7). In conjunction with the known irritant properties of BPA (4), our interpretation of these results is that even BPA has only marginal biocompatibility. In spite of the lack of a strong inflammatory tissue response in preliminary, subcutaneous implantation studies, we currently suggest that BPA-containing, degradable polymers may be of only limited value as biomaterials.

In an attempt to identify more biocompatible diphenols for the design of degradable biomaterials, we studied derivatives of tyrosine dipeptide as potential monomers. After protection of the amino terminus and the carboxylic acid terminus, the reactivity of tyrosine dipeptide (Figure 1) could be expected to be similar to the reactivity of industrial diphenols. Thus, derivatives of tyrosine dipeptide could be suitable replacements for BPA in the synthesis of a variety of new polymers that had heretofore not been accessible as biomaterials due to the lack of diphenolic monomers with good biocompatibility.

In this context, we were particularly interested in the exploration of polyiminocarbonates (2,8) and polycarbonates (3), two structurally related classes of polymers that are best derived from diphenolic monomers. The corresponding BPA-derived polymers (Figure 2) had previously been found to be particularly strong materials with excellent engineering properties (2,3). In order to identify tyrosine derivatives that could be used in place of BPA, we initiated a detailed investigation of the structure-property relationships in polyiminocarbonates (9,10) and polycarbonates (11). These studies led to the identification of desaminotyrosyl-tyrosine hexyl ester (Dat-Tyr-Hex, also abbreviated as DTH) as the first tyrosine derived monomer whose polymerization leads to strong and processible polymers with a wide range of potential biomedical applications.

Experimental

All solvents were of HPLC grade and were used as received, except for tetrahydrofuran (THF) which was freshly distilled from sodium and benzophenone prior to use.

Molecular Weight. Molecular weights were determined by gel permeation chromatography (GPC) on a chromatographic system consisting of a Perkin-Elmer Model 410 pump, a Waters Model 410 RI detector, and a Perkin-Elmer Model 3600 computerized data station. For polymer analysis, two PL-gel GPC columns (300 mm x 7.7 mm, particle size 5 μm, pore size 10^5 and 10^3 Å respectively) were placed in series and were operated at a flow rate of 1 mL/min in DMF containing 0.1% (w/v) of LiBr. Molecular weights were calculated relative to polystyrene standards without further corrections.

Thermal Properties. The glass transition temperature (T_g) and the decomposition temperature (T_d) were measured with a DuPont 910 Differential Scanning Calorimeter (DSC) calibrated with indium. The standard heating rate for all polymers was 10 °C/min. Thermogravimetric analysis (TGA) was performed on a DuPont 951 Thermogravimetric Analyzer at a heating rate of 20 °C/min.

Polymer Processing. Polymer films were cast in trimethylsilyl coated glass molds from membrane filtered 15% (w/v) methylene chloride or chloroform solutions. Transparent films were obtained which were dried to constant weight in high vacuum. Rectangular strips or round disks were cut from the films. For compression molding a Carver laboratory press equipped with thermostated, heated platens was used. Polymers were placed in a stainless steel mold and heated to 40 °C above their glass transition temperature. Then a load of 1-2 tons was applied for 5 min.

Mechanical Properties. The mechanical properties of thin, solvent-cast polymer films were measured on an Instron Tensile Tester according to ASTM standard D882-83. In all cases, tensile values were calculated from the average of at least four measurements obtained from four separate specimens per polymer sample.

Hydrolytic Degradation Studies. Disks (2.3 cm x 1.1 cm x 0.05 cm, approximately 150 mg) were cut from solvent-cast films. The disks were incubated at 37°C in phosphate buffer (0.1 M, pH 7.4). The degradation process was followed by recording the weight change of individual disks, by measuring the residual polymer molecular weight after various intervals of exposure to the buffer solution and by FT-IR analysis of partly degraded samples.

Synthesis of Tyrosine Derivatives. Tyrosine derived monomers were prepared by DCC mediated coupling reactions in THF following standard procedures of peptide chemistry (*12*). Dat-Tym, Z-Tyr-Tym, Z-Tyr-Tyr-Hex, and Dat-Tyr-Hex (DTH) were purified and characterized according to reference 10. For spectral data of Z-Tyr-Tyr-Hex, see Kohn and Langer (*13*).

Synthesis of Dicyanates. All dicyanates were prepared according to previously published procedures (*2*). For IR spectral data see Kohn and Langer (*2*), for elemental analyses, see reference 10. A specific procedure for the preparation of Z-Tyr-Tyr-Hex-dicyanate has been published (*13*).

Preparation of Polyiminocarbonates. Solution polymerizations of equimolar mixtures of diphenols and dicyanates were carried out in THF using potassium tert-butoxide as the catalyst as described previously (*8*). Interfacial polymerizations were performed in a two phase system consisting of aqueous NaOH and methylene chloride in the presence of 10% (molar) of tetrabutylammonium bromide as a phase transfer catalyst (*8*). All polyiminocarbonates showed a strong and sharp IR absorption band at 1670 cm^{-1} to 1680 cm^{-1}, which is characteristic of the iminocarbonate bond.

Preparation of Polycarbonates. About 30 mmoles of a tyrosine derived diphenolic monomer was placed into a 250 mL flask. After addition of 75 mL of dry methylene chloride and 11 mL of anhydrous pyridine, a pale yellow solution was obtained. At room temperature, a molar excess of phosgene (as a 1.94 M solution in toluene) was added slowly to the vigorously stirred solution over a period of about 90 minutes. The molecular weight of the polymeric product was monitored by GPC and the addition of phosgene was terminated after the maximum molecular weight had been obtained. Stirring was continued for an additional 120 minutes. Thereafter, the reaction mixture was diluted with 500 mL of methylene chloride, transferred into a separatory funnel and extensively extracted with 0.2 N aqueous HCL. The organic phase (containing the polymer) was then dried over magnesium sulfate and concentrated to 150 mL. The polymer was precipitated by slowly adding the concentrated solution into 750 mL of hexane. Yield: about 85%; intrinsic viscosity (chloroform, 30°C): 0.6 to 1.0 dL/g, depending on monomer purity.

Results and Discussion

In an attempt to identify new, biocompatible diphenols for the synthesis of polyiminocarbonates and polycarbonates, we considered derivatives of tyrosine dipeptide as potential monomers. Our experimental rationale was based on the assumption that a diphenol derived from natural amino acids may be less toxic than many of the industrial diphenols. After protection of the amino and carboxylic acid groups, we expected the dipeptide to be chemically equivalent to conventional diphenols. In preliminary studies (*14*) this hypothesis was confirmed by the successful preparation of poly(Z-Tyr-Tyr-Et iminocarbonate) from the protected tyrosine dipeptide Z-Tyr-Tyr-Et (Figure 3). Unfortunately, poly(Z-Tyr-Tyr-Et iminocarbonate) was an insoluble, nonprocessible material for which no practical applications could be identified. This result illustrated the difficulty of balancing the requirement for biocompatibility with the need to obtain a material with suitable "engineering" properties.

In order to identify tyrosine derivatives that would lead to polymers that are processible, mechanically strong, and also biocompatible, we initiated a detailed investigation of the structure-property relationships in polyiminocarbonates and polycarbonates. Since the amino and carboxylic acid groups of tyrosine dipeptide (the N and C termini) provide convenient attachment points, selected pendent chains can be used to modify the overall properties of the polymers. This is an important structural feature of tyrosine dipeptide derived polymers.

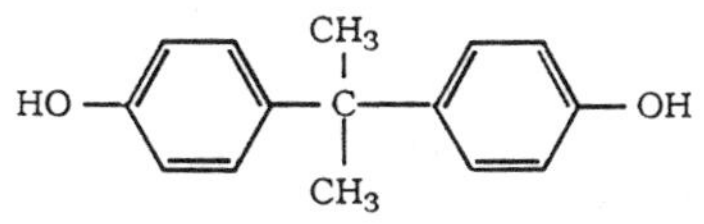

Bisphenol A

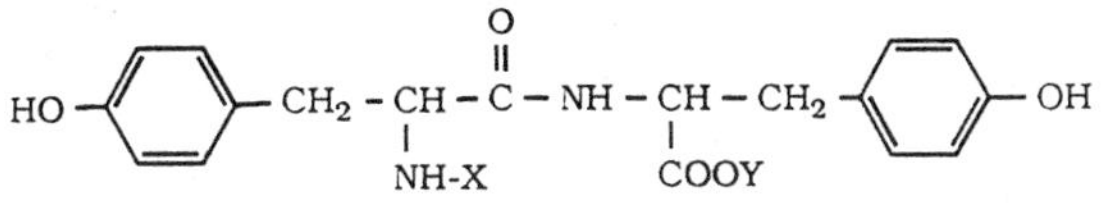

protected tyrosine dipeptide

Figure 1. Molecular structures of Bisphenol A and fully protected tyrosine dipeptide. The amino and carboxylic acid groups of the dipeptide are rendered unreactive by protecting groups (schematically represented by X and Y). This leaves the phenolic hydroxyl groups as the only reactive sites of the molecule.

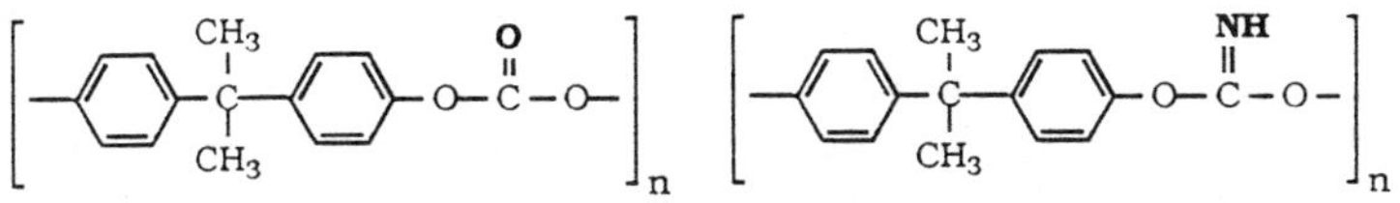

poly(BPA-carbonate)   poly(BPA-iminocarbonate)

Figure 2. Molecular structures of poly(Bisphenol A carbonate) and poly(Bisphenol A iminocarbonate).

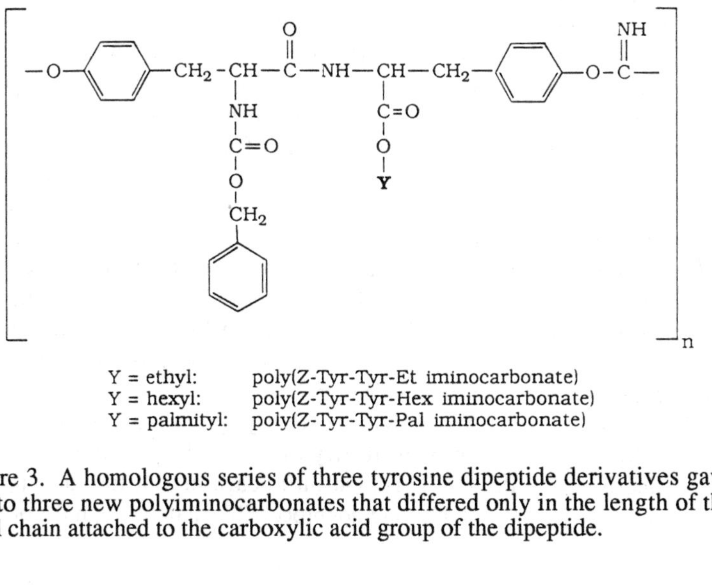

Y = ethyl: poly(Z-Tyr-Tyr-Et iminocarbonate)
Y = hexyl: poly(Z-Tyr-Tyr-Hex iminocarbonate)
Y = palmityl: poly(Z-Tyr-Tyr-Pal iminocarbonate)

Figure 3. A homologous series of three tyrosine dipeptide derivatives gave rise to three new polyiminocarbonates that differed only in the length of the alkyl chain attached to the carboxylic acid group of the dipeptide.

The Effect of Pendent Chains on the Physicomechanical Properties of Tyrosine Derived Polymers. In order to investigate the influence of the C-terminus protecting groups on the polymer properties, we prepared the ethyl, hexyl and palmityl esters of N-benzyloxycarbonyl-L-tyrosyl-L-tyrosine (15). These monomers represented a homologous series which made it possible to prepare three new polyiminocarbonates differing only in the length of the alkyl group attached to the C-terminus (Figure 3).

The first member of this series, poly(Z-Tyr-Tyr-Et iminocarbonate), was a virtually insoluble polymer that was not processible. Poly(Z-Tyr-Tyr-Hex iminocarbonate), the second member of the series, was soluble in many common organic solvents such as chlorinated hydrocarbons, THF, DMF, and DMSO. Solvent casting yielded transparent, but brittle films. The increasing length of the pendent alkyl group also reduced the polymer glass transition temperature (T_g) by 29 °C (Table I). Poly(Z-Tyr-Tyr-Pal iminocarbonate), the third member of the series, was very similar to poly(Z-Tyr-Tyr-Hex iminocarbonate) in spite of a large increase in the length of the C-terminus protecting group from C_6 to C_{16}. These observations indicate that the C-terminus protecting group must have a minimum length in order for the tyrosine derived polyiminocarbonate to become soluble in organic solvents. However, increasing the length of the C-terminus alkyl ester chain beyond C_6 did not have any further advantage. We therefore selected the hexyl ester as the standard C-terminus protecting group throughout the remainder of our studies. Currently, a more detailed investigation of the relationship between solubility, T_g, and mechanical strength on one hand and the length of the ester side chain on the other hand is in progress.

In a second series of experiments (9,10), tyrosine derived dipeptides were synthesized in which the amino group and/or the carboxylic acid group of tyrosine were replaced by hydrogen atoms. This was achieved by using the tyrosine derivatives desaminotyrosine (Dat), and tyramine (Tym) instead of tyrosine (Figure 4). Dat occurs naturally in plants and Tym is a human metabolite. Using these building blocks, four structurally related dipeptide derivatives were prepared that carried either no pendent chains at all (Dat-Tym), a N-benzyloxycarbonyl group as the only pendent chain (Z-Tyr-Tym), a hexyl ester group as the only pendent chain (Dat-Tyr-Hex, further abbreviated as DTH), or both types of pendent chains simultaneously (Z-Tyr-Tyr-Hex). This series of monomers (Figure 4) made it possible to investigate the contribution of each type of pendent chain separately. Of course, Z-Tyr-Tyr-Hex is identical to the monomer used before (Figure 3).

The corresponding polyiminocarbonates (Figure 5) were prepared first, using recently developed polymerization procedures (8). Poly(Dat-Tym iminocarbonate), the polymer carrying no pendent chains at all, was an insoluble material. Thermal processing techniques could not be used due to the low thermal stability of the polymer in the molten state. Thus poly(Dat-Tyr iminocarbonate) was a virtually non-processible material without practical applications.

The presence of a stiff benzyloxycarbonyl group in Z-Tyr-Tym did not significantly improve the processibility or solubility of the corresponding polyiminocarbonate: Poly(Z-Tyr-Tym iminocarbonate), like poly(Dat-Tym iminocarbonate), was an insoluble, and nonprocessible material. On the other hand,

Table I. Physicomechanical Properties of Selected Polymers

Polymer [molecular weight][a]	Glass Transition T_g (°C)	Decomposition onset (°C)	Decomposition peak (°C)	Tensile strength (kg/cm²)	Tensile modulus (kg/cm²)	Elongation at yield (%)	Elongation at break (%)	Solubility[b]
Poly(BPA iminocarbonate)[109,000]	69	151	171	510	22,000	3.5	4.0	yes
Poly(Z-Tyr-Tyr-Et iminocarbonate)[n/a][e]	91	145	163	-	-	-	-	no[c]
Poly(Z-Tyr-Tyr-Hex iminocarbonate)[54,000]	62	143	160	-	-	-	-	yes
Poly(Z-Tyr-Tyr-Pal iminocarbonate)[69,000]	60	141	163	-	-	-	-	limited[d]
Poly(Dat-Tym iminocarbonate)[71,500]	not observed[f]	130	147	-	-	-	-	limited[d]
Poly(Dat-Tyr-Hex iminocarbonate)[103,000]	55	158	175	400	16,300	3.5	7.5	yes
Poly(BPA carbonate)[109,000]	149	416	486	625	22,000	6	60	yes
Poly(Dat-Tyr-Hex carbonate)[262,000]	62	305	355	335	13,900	3	100	yes
Poly(Z-Tyr-Tyr-Hex carbonate)[400,000]	86	232	261	340	13,200	4	5	yes

a weight average molecular weight as determined by GPC relative to polystyrene standards without further correction.

b polymers were classified as soluble if clear solutions in chlorinated hydrocarbons, DMF, THF, and DMSO were obtained at a concentration of at least 10 mg/mL.

c no solvent could be identified.

d limited solubility in DMF/LiBr or DMF/chloroform mixtures only.

e due to the insoluble nature of the polymer, the molecular weight could not be determined.

f no glass transition was observed from 0°C up to the onset of thermal decomposition.

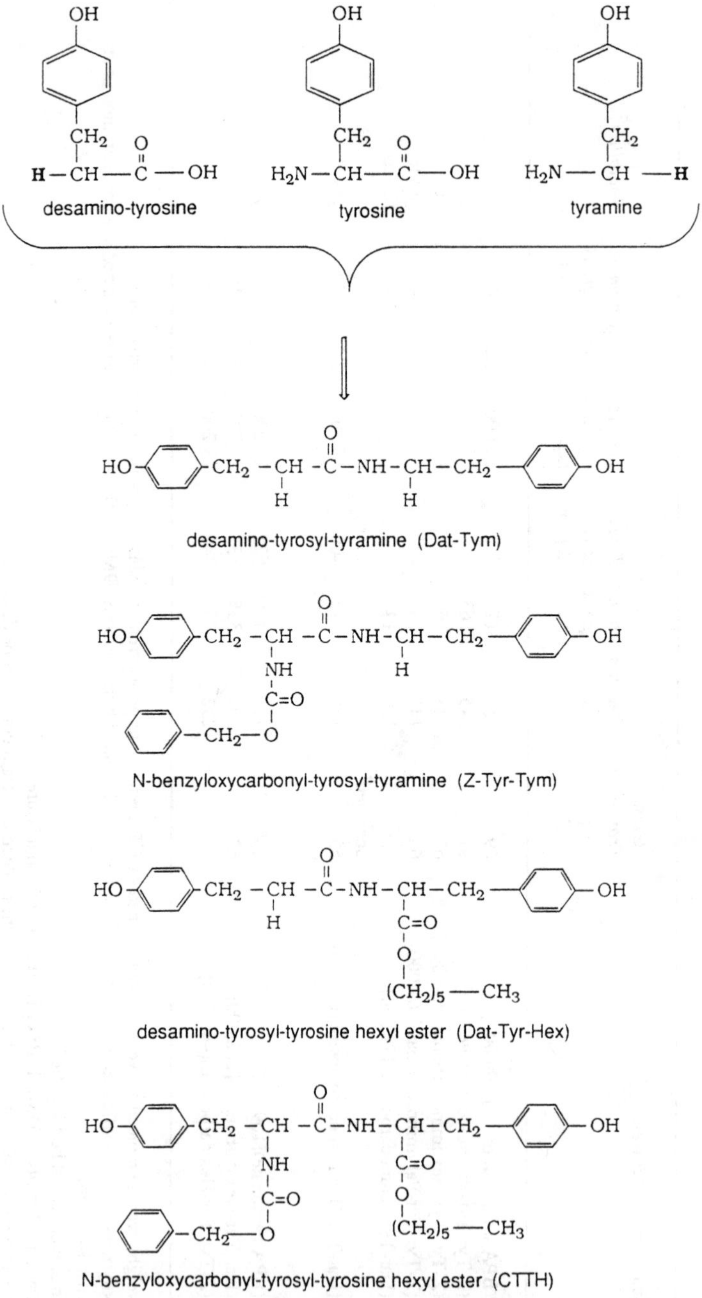

desamino-tyrosine   tyrosine   tyramine

desamino-tyrosyl-tyramine (Dat-Tym)

N-benzyloxycarbonyl-tyrosyl-tyramine (Z-Tyr-Tym)

desamino-tyrosyl-tyrosine hexyl ester (Dat-Tyr-Hex)

N-benzyloxycarbonyl-tyrosyl-tyrosine hexyl ester (CTTH)

Figure 4. Three naturally occurring tyrosine derivatives (desaminotyrosine, tyrosine, and tyramine) were used to prepare four different monomeric diphenols that carry no pendent chains (Dat-Tym), only a benzyloxycarbonyl (Z) group, only a hexyl ester group, or both types of pendent chains (Z-Tyr-Tyr-Hex).

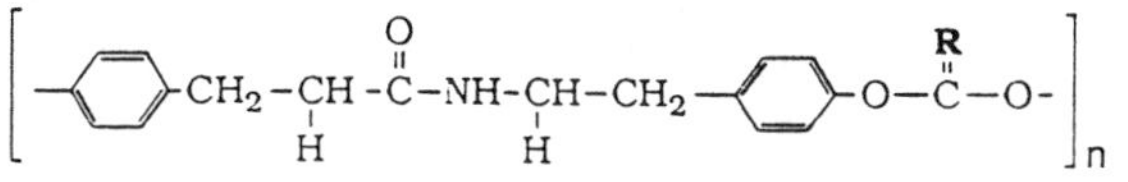

R = NH: poly(Dat-Tym iminocarbonate)
R = O: poly(Dat-Tym carbonate)

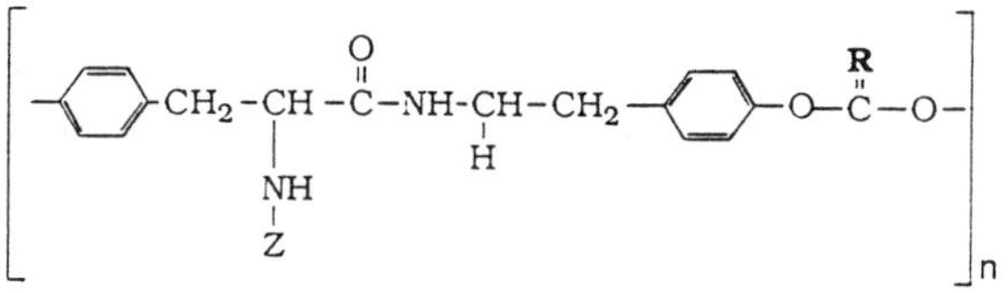

R = NH: poly(Z-Tyr-Tym iminocarbonate)
R = O: poly(Z-Tyr-Tym carbonate)

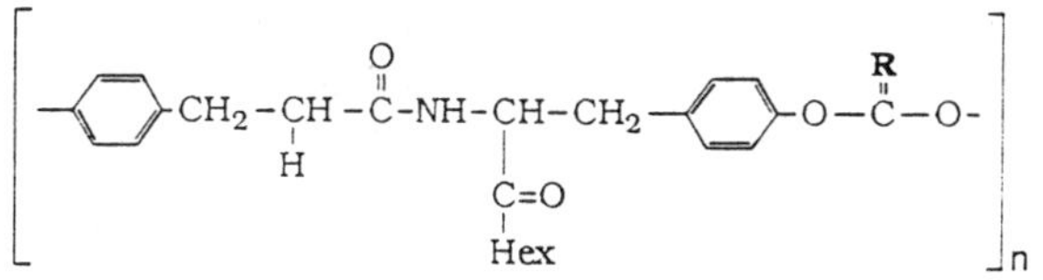

R = NH: poly(Dat-Tyr-Hex iminocarbonate)
R = O: poly(Dat-Tyr-Hex carbonate)

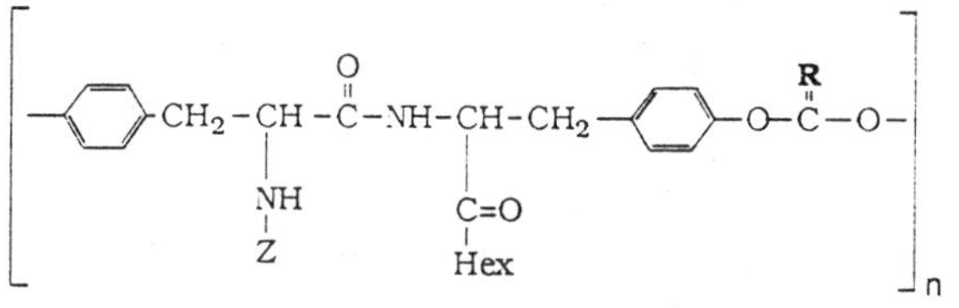

R = NH: poly(Z-Tyr-Tyr-Hex iminocarbonate)
R = O: poly(Z-Tyr-Tyr-Hex carbonate)

Figure 5. Each monomer shown in Figure 4 was used in the synthesis of a polyiminocarbonate and a polycarbonate, resulting in the preparation of eight structurally related polymers.

the presence of a flexible hexyl ester group in Dat-Tyr-Hex (DTH) improved the physicomechanical properties of the resulting polymer significantly: among all tested polyiminocarbonates, poly(Dat-Tyr-Hex iminocarbonate) was the most soluble material and had the lowest glass transition temperature (55°C), (Table I).

The softening point of poly(Dat-Tyr-Hex iminocarbonate) was 70°C, making it possible to process this polymer by compression molding at temperatures ranging from about 70 to 80°C. In this way, clear and mechanically strong disks were readily obtained. The low processing temperature of poly(Dat-Tyr-Hex iminocarbonate) is a particular advantage in the fabrication of drug delivery devices since even relatively heat sensitive drugs can be safely incorporated into the polymeric matrix.

Among the polyiminocarbonates, only poly(Z-Tyr-Tyr-Hex iminocarbonate) and poly(Dat-Tyr-Hex iminocarbonate) were soluble enough to be processible by solvent casting. While poly(Z-Tyr-Tyr-Hex iminocarbonate) formed brittle films that could not be handled easily without breaking, poly(Dat-Tyr-Hex iminocarbonate) formed tough films. These films had a tensile strength of about 400 kg/cm^2 and ranked among the strongest, degradable implant materials currently available (1). Thus, the most promising polyiminocarbonate was obtained when Dat-Tyr-Hex (DTH) was used as the monomeric starting material.

Next we used the same four monomers for the preparation of the corresponding polycarbonates (11). The physicomechanical properties of the tyrosine derived polycarbonates (Figure 5) followed the same basic pattern as observed for the polyiminocarbonates: poly(Dat-Tym carbonate), the polymer carrying no pendent side chains, was the least soluble of all four polycarbonates and dissolved at room temperature only in N-methyl pyrrolidone. The introduction of the benzyloxycarbonyl group had no significant effect on the polymer solubility, while the introduction of the hexyl ester group led to poly(Dat-Tyr-Hex carbonate), a polymer that was freely soluble in a large variety of organic solvents and showed excellent processibility in a variety of molding techniques. The combination of two types of side chains in Z-Tyr-Tyr-Hex provided no advantage, as shown by the reduced solubility of poly(Z-Tyr-Tyr-Hex carbonate) as compared to poly(Dat-Tyr-Hex carbonate). Furthermore, films of poly(Z-Tyr-Tyr-Hex carbonate) obtained by solvent casting were brittle, while films of poly(Dat-Tyr-Hex carbonate) were ductile and tough and failed only at an elongation of about 100%. Again, the most promising polymer was obtained when Dat-Tyr-Hex (DTH) was used as the monomeric starting material.

Based on these observations we concluded that similar structure-property relationships govern the physicomechanical properties of both polyiminocarbonates and polycarbonates. Polymers derived from Dat-Tym (carrying no pendent side chains) were insoluble and, clearly, the presence of pendent chains is an important structural requirement. There is, however, a very significant difference between the effect of a stiff benzyloxycarbonyl group, attached to the *N*-terminus, and the effect of a flexible alkyl group, attached to the *C*-terminus of the dipeptide. The introduction of a hexyl ester group appeared to be particularly effective in increasing the solubility of the polymer and reducing its softening temperature, two important parameters that determine the overall processibility of a polymeric material. The combination of both types of pendent chains within the same polymer had no advantage. Polymers derived from Z-Tyr-Tyr-Hex were no better in terms of their solubility, processibility, and mechanical strength than polymers derived from Dat-Tyr-Hex.

A Comparison of Polyiminocarbonates and Polycarbonates: The Effect of the NH Group on Polymer Properties. The replacement of the carbonyl oxygen by an NH group presents the only molecular difference between polyiminocarbonates and polycarbonates. In spite of the overall structural similarity between these two types of polymers, we found very significant differences between their respective material properties. In general, polyiminocarbonates and polycarbonates tend to complement each other in several aspects.

Solvent-cast films of tyrosine-derived polyiminocarbonates and polycarbonates were virtually indistinguishable in appearance and exhibited similar morphological features. When examined by X-ray diffraction, using an automated Scintag Pad V diffractometer all tested films were found to be completely amorphous. The lack of ordered domains in solvent-cast films seemed to be a general feature of tyrosine-derived polyiminocarbonates and polycarbonates, irrespective of the pendent chain configuration of the monomeric dipeptide.

The tensile properties (tensile strength, Young's modulus, and elongation) of unoriented, noncrystalline films were investigated for those polymers that formed clear films by solvent casting. The results collected in Table I allowed several general conclusions.

Polyiminocarbonates and polycarbonates tend to be high strength materials. Polyiminocarbonates were somewhat stronger than the corresponding polycarbonates. Typical values for the tensile strength of poly(Dat-Tyr-Hex iminocarbonate) were about 400 to 500 kg/cm^2 (40 to 50 MPa), while the corresponding poly(Dat-Tyr-Hex carbonate) had a tensile strength of about 340 kg/cm^2 (34 MPa). These values compare favorably to the tensile strength values obtained for some of the strongest degradable implant material currently available e.g., poly(L-lactic acid) (*1*). For comparison, for wet films of poly(γ-methyl-D-glutamate) a much lower tensile strength of only 50 to 100 kg/cm^2 (5 to 10 MPa) had been reported (*16*).

Whereas the tensile strength was not a sensitive function of the monomer structure, the tensile modulus (Young's Modulus) was clearly related to the monomer structure. This is expected since the tensile modulus is a measure of the polymer's resistance to deformation and is related to the "stiffness" of a polymeric material. The highest tensile modulus (22,000 kg/cm^2, 2.2 GPa) was measured for poly(BPA iminocarbonate). Replacement of BPA by Dat-Tyr-Hex reduced the tensile modulus significantly. This observation can possibly be attributed to the presence of the long hexyl ester pendent chain in Dat-Tyr-Hex. Generally, the polyiminocarbonates were somewhat "stiffer" than the corresponding polycarbonates. Thus, the tensile moduli of poly(Dat-Tyr-Hex iminocarbonate) and poly(Dat-Tyr-Hex carbonate) were 16,300 kg/cm^2 (1.6 GPa) and 13,900 kg/cm^2 (1.3 GPa) respectively.

The different degree of ductility represented one of the most striking differences between polyiminocarbonates and polycarbonates. Polycarbonates were usually ductile and tough materials, often failing only at about 100% of elongation while all tested polyiminocarbonates were brittle materials failing at about 2 to 7% of elongation. This difference was related to the backbone linkages and not to the molecular structure of the monomers as demonstrated by the fact that poly(Dat-Tyr-Hex iminocarbonate) and poly(Dat-Tyr-Hex carbonate) failed at 7% and 100% elongation respectively, despite being derived from identical monomers.

The degree of hydrolytic stability is another important difference between polyiminocarbonates and polycarbonates. The chemical mechanism of the hydrolytic degradation of polyiminocarbonates had been studied previously (2) and had been shown to lead to the formation of ammonia, carbon dioxide and the regeneration of the diphenol used in the synthesis of the polymer. An important side reaction during the degradation process was the formation of carbonate linkages whose presence could be readily ascertained by IR spectroscopy of partially degraded polyiminocarbonates (2).

After exposure of solvent-cast films to aqueous buffer solutions, the molecular weight of all tyrosine-derived polyiminocarbonates decreased rapidly to about 1,000 to 6,000 Daltons (weight average) resulting in a concomitant loss of mechanical strength. The films became turbid within a matter of days and swelled noticeably due to the absorption of water. Thus, the iminocarbonate bond was found to be surprisingly unstable and polyiminocarbonates are among the most rapidly degrading biomaterials currently under investigation.

The fast, initial degradation process slowed down considerably once the molecular weight had decreased to about 1,000 to 6,000 Daltons. At this point, the residues consisted mostly of carbonate containing oligomers. Under simulated physiological conditions in vitro, the complete erosion of these carbonate containing residues required approximately one year. The in vitro erosion of polyiminocarbonates is therefore best described in terms of a fast, initial phase, followed by a slow second phase (10).

On the other hand, the hydrolytic degradation of the tyrosine derived polycarbonates was very slow. Contrary to the polyiminocarbonates, polycarbonates absorbed very little water and the molecular weight of poly(Dat-Tyr-Hex carbonate) decreased only slowly with time. Exposure of thin films to phosphate buffer solution at pH 7.4 (37°C) resulted in a 50% reduction of the initial molecular weight of poly(Dat-Tyr-Hex carbonate) over 6 months and in a 75% reduction over one year. A useful feature of the degradation process of poly(Dat-Tyr-Hex carbonate) is that the films retained their physical integrity throughout the one year degradation study. Contrary to the degradation process observed for poly(lactic acid) and many other polymers, thin films of poly(Dat-Tyr-Hex carbonate) did not crumble or disintegrate over a one year period, although they lost much of their initial mechanical strength.

When films of poly(Dat-Tyr-Hex carbonate) were exposed to strongly alkaline conditions (1 M NaOH at 37°C), a rapid decrease in the weight of the devices was seen. The thickness of the films decreased steadily, while the molecular weight of the intact core of the polymer matrix remained virtually unchanged. These observations indicated that the polymer degraded by surface erosion. However, since the erosion profile in 1 M NaOH is of little practical importance, this phenomenon was not further investigated.

Throughout the hydrolysis of poly(Dat-Tyr-Hex carbonate), free hexanol (easily identified by its characteristic smell or by GC) was released into the hydrolysis medium. We currently assume that the hydrolysis of the ester function is one of the first steps of the chemical degradation mechanism of poly(Dat-Tyr-Hex carbonate). Hydrolysis of the side chain ester functions would result in the formation of unprotected carboxylic acid groups which could contribute to the hydrolytic degradation of the polymer backbone, either by a general acid catalysis effect on the hydrolysis of the carbonate or amide backbone linkages or by partial solubilization of the polymer chains. This mechanism would explain why poly(Dat-Tyr-Hex carbonate) degrades at all, while the widely used poly(BPA carbonate) is known to be virtually nondegradable under physiological conditions.

Biocompatibility of Dat-Tyr-Hex. An evaluation of the tissue compatibility of Dat-Tyr-Hex derived polyiminocarbonate and polycarbonate in rats has recently been concluded (5). These studies included medical grade poly(L-lactic acid) as an internal standard, medical grade polyethylene as a negative control, and poly(BPA iminocarbonate) as a positive control. At times ranging from 7 days to 4 months post implantation, the subcutaneous implantation sites were histologically evaluated. Based on this experimental design, it was possible to compare the tissue response observed for tyrosine derived polymers relative to polyethylene (mild response) and poly(L-lactic acid) (moderate response). The general conclusion was that the tissue response elicited by the slowly degrading poly(Dat-Tyr-Hex carbonate) was comparable to the mild response seen for polyethylene, while the fast degrading poly(Dat-Tyr-Hex iminocarbonate) elicited a response that was comparable to the tissue response seen for medical grade poly(L-lactic acid). On the other hand, poly(BPA iminocarbonate) elicited an inflammatory response that was clearly more severe (as judged by the number and density of inflammatory cells at the implantation site) than the responses observed for any of the other polymers.

A more comprehensive biocompatibility evaluation of Dat-Tyr-Hex derived polyiminocarbonates and polycarbonates is currently in progress, including an evaluation of the antigenicity, thrombogenicity, and toxicity of Dat-Tyr-Hex and its polymers. In particular, the in vivo fate of the low molecular weight oligomers that are formed during the degradation of polyiminocarbonates is still under investigation.

Conclusions

Because of the limited biocompatibility of diphenols, these versatile and valuable monomers had not been used in the design of degradable biomaterials. Our initial attempt to utilize Z-Tyr-Tyr-Et (a protected tyrosine dipeptide, Figure 3) as a monomer for the preparation of degradable biomaterials yielded polymers with highly unfavorable engineering properties. This result illustrated the difficulty in balancing the requirement for nontoxicity with the requirement for good engineering properties. We therefore established general structure-property relationships that made it possible to identify desaminotyrosyl-tyrosine alkyl esters as promising monomers for the synthesis of new biomaterials containing aromatic backbone structures. So far, we have identified desaminotyrosyl-tyrosine hexyl ester (Dat-Tyr-Hex or DTH) as a particularly useful monomer. Poly(Dat-Tyr-Hex iminocarbonate) and poly(Dat-Tyr-Hex carbonate) were strong and readily processible materials with excellent engineering properties. These materials are the first tyrosine derived polymers with a wide range of potential biomedical applications.

Poly(Dat-Tyr-Hex iminocarbonate) is an amorphous polymers that exhibits a high degree of mechanical strength and stiffness. This material may be suitable for applications that require a temporary, mechanical support function. Noteworthy is the rapid degradation (several days) of this polymer to low molecular weight oligomers. Possible limitations of the practical applicability of poly(Dat-Tyr-Hex iminocarbonate) as a biomaterial are the low thermal stability of the iminocarbonate linkage and the complicated, two phase degradation mechanism that leads to the formation of slowly degrading residues of low molecular weight.

Poly(Dat-Tyr-Hex carbonate) is a slowly degrading, strong and ductile material. This material may be applicable in situations where a mechanical support function is needed for an extended period of time such as in small bone fixation devices (bone pins or screws). The reproducible fabrication of such devices by injection molding is particularly easy due to the amorphous character of poly(Dat-Try-Hex carbonate).

Initial tests in the rat revealed a high degree of tissue compatibility of Dat-Tyr-Hex derived polymers. More detailed tests are now in progress. In addition, tyrosine derived polymers are currently being evaluated in the formulation of an intracranial controlled release device for the release of dopamine, in the design of an intraarterial stent (to prevent the restenosis of coronary arteries after balloon angioplasty), and in the development of orthopedic implants. The use of tyrosine derived polymers in these applications will provide additional data on the biocompatibility of these polymers.

Acknowledgment

The author acknowledges the assistance of Mr. Israel Engelberg, who performed some of the mechanical tests. The X-ray diffraction data were collected in the laboratory of Professor M. Greenblatt (Department of Chemistry, Rutgers University) by Ms. Aruna Nathan. The author thanks Mr. Chun Li and Mr. Satish Pulapura for their part in the synthesis and characterization of tyrosine derived polymers.

Literature Cited

1.   Engelberg, I.; Kohn, J. *Biomaterials* **1991**, in press.
2.   Kohn, J.; Langer, R. *Biomaterials* **1986**, *7*, 176.

3. Schnell, H. *Chemistry and Physics of Polycarbonates,* Interscience Publishers, John Wiley & Sons, New York, NY, 1964.
4. National Institute for Occupational Safety and Health (NIOSH), Registry of Toxic Effects of Chemical Substances, 1981-82, p.70.
5. Silver, F.H., Marks, M., Kato, Y. P., Li, C., Pulapura, S., and Kohn, J., Biocompatibility of Degradable Polycarbonates and Polyiminocarbonates, manuscript in preparation.
6. Leong, K., private communication.
7. Kohn, J. and Silver, F. H. unpublished results.
8. Li, C.; Kohn, J. *Macromolecules* **1989**, *22,* 2029.
9. Pulapura, S.; Kohn, J. *Polym. Preprints* **1990**, *31* (1), 233.
10. Pulapura, S.; Li, C.; Kohn, J. *Biomaterials* **1990**, *11*, 665.
11. Pulapura, S. and Kohn, J. Tyrosine Derived Polycarbonates: New Polymers for Medical Applications, manuscript in preparation.
12. Bodanszky, M. and Bodanszky, A. *The Practice of Peptide Synthesis;* Springer Verlag, New York, 1984, p. 145.
13. Kohn, J., Langer, R. *J. Am. Chem. Soc.* **1987**, *109,* 817.
14. Kohn, J. and Langer, R. A New Approach to the Development of Bioerodible Polymers for Controlled Release Applications Employing Naturally Occurring Amino Acids, In *Proceedings of the ACS Division of Polymeric Materials: Science and Engineering,* American Chemical Society, **1984**, *51,* 119.
15. Kohn, J. and Langer, R. In *Peptides, Chemistry and Biology,* Marshall, G. R., Ed.; Escom Publishing: Leiden, Netherlands, 1988, pp. 658-661.
16. Mohadger, Y.; Wilkes, G. L. *J. Poly. Sci. Poly. Phys. Ed.* **1976**, *14,* 963.

RECEIVED March 26, 1991

Chapter 16

Kinetics of Controlled Release
from Acid-Catalyzed Matrices

Abhay Joshi and Kenneth J. Himmelstein[1]

Allergan Pharmaceuticals, 2525 Dupont Drive, Irvine, CA 92715—1599

A mathematical model for the analysis of basic physicochemical determinants that yields experimentally verifiable predictions of erosion characteristics and controlled release of bioactive agents from bioerodible matrices is presented. There is considerable experimental information in the literature on the detailed performance of the poly(ortho ester) system, covering a wide range of system characteristics, and therefore making a good test case for mathematical analysis. The analysis shows that the dynamic changes in polymer matrix properties, namely, simultaneous reaction-diffusion-transport of matrix constituents, moving diffusion and water front, water-polymer partition coefficients, solubility of water and diffusivity of matrix components as a function of extent of acid catalyzed matrix hydrolysis, play a significant role in regulating the release kinetics of bioactive agents. The model predicts experimentally measureable quantities: release characteristics of incorporated bioactive agents, water penetration into the matrix, and catalytic degradation of the polymer matrix. Further, a good estimate of the concentration of unbroken polymer backbone linkages and hence the molecular weight of the polymer disc with time is obtained using random scission kinetics.

Drug release in a controlled fashion for a variety of applications from a single generalized system has long been a substantial goal of drug delivery research. One of the approaches taken to achieve this goal has been the use of hydrolytically unstable polymers which contain drug. To achieve a wide range of delivery durations in these systems, the rate of hydrolysis can be altered by choice of copolymer to give intrinsically different reaction rates and (by virtue of different glass transition

[1]Current address: Himmelstein and Associates, Pearl River, NY 10965

0097—6156/91/0469—0170$06.00/0
© 1991 American Chemical Society

temperatures) different water permeation rates, and by including erosion accelerating acid groups or core stabilizing basic excipients. It is desirable that the rate of drug release from these systems be controlled by the hydrolysis of the polymer itself. The kinetics of reaction, however, depend on many different physical and chemical processes that occur in tandem: diffusion of water into the matrix, reaction with the polymer, physical degradation of the polymer, diffusion of the drug through the degrading matrix, and secondary processes such as the production and diffusion of secondary molecules like catalysts and stabilizers. In order to assess the effect of all these interrelated processes on the performance of these eroding polymeric release systems, a means of systematically studying them is required. Therefore, a mathematical model which includes all of the important reaction and transport properties in a mechanistic fashion can be a useful tool to study the performance of these complex systems.

A previous model (*1*) was developed for this purpose. This model was able to predict the general performance characteristics of a polymeric delivery system which used a small diffusible molecule to act as a catalyst to accelerate the erosion rate. However, this model was never applied to the simulation of experimental data using independently measured parameters such as rate constants and diffusivities to determine its quantitative predictive capabilities. Secondly, to determine whether or not a system is really performing as believed, a more rigorous method than simply calculating the rate of drug release should be employed, since many different sets of processes can lead to very similar release characteristics. The previously developed model had as its major dependent variable, the concentration of hydrolyzable polymeric linkages at any position and time. While this variable was useful for simulations, it is not an easily verifiable measure to describe experimental systems. Thus in this paper, the model is extended to the calculation of polymer molecular weight distribution as a function of time and position using random degradation kinetics. This variable is then used to calculate the average molecular weight for these systems. In addition, the model is updated to include a position dependent water partition coefficient at the model-defined polymer-environmental barrier. Finally, it is desirable to relate gross erosion characteristics - bulk or surface erosion - to an easily defined characteristic of the system. The Thiele modulus, an indicator that determines the regimes of diffusion and reaction controlled kinetics of the system, is used to describe the overall system performance.

In this paper, then, the previously developed model (*1*) is extended to the calculation of erosion characteristics of a well described polymeric delivery system, the acid-catalyzed erosion of poly(ortho ester)s (*2-6*). This system is chosen as the example system because of the completeness of the data package in the open literature. It is expected that this modelling approach is also useful for other hydrolytically unstable polymeric drug delivery systems.

Experimental

Model Structure. As one of the main goals of the present effort is to use the extended mathematical analysis to model experimentally described results, the model was set up to simulate the physical system studied by Nguyen, et al. (2). In these experiments, an acid producing species, phthalic anhydride, is mixed into the poly(ortho ester) matrix, along with the drug to be released. The general system is shown in Figure 1. In operation, water (A) from the environment diffuses into the poly(ortho ester) disc to react with the acid producing species (B) to produce the acid catalyst (C). The catalyst is free to react with the hydrolytically unstable polymer linkage (D) to form an activated unstable intermediate complex (D*) which then can further react with water (A) to break the backbone linkages. As the polymer erodes, there is a subsequent loss of physical integrity of the polymeric system and drug (E), along with small degradation products, is free to diffuse out.

The rate of drug release (E) from the eroding matrix is controlled by: (a) the chemical properties of the system - the hydrolytic and the neutralizing process at the boundary of the device, catalytic degradation of the polymer and the intrinsic backbone reactivity, and (b) several concomitant physical processes such as water diffusivity, water solubility, water partitioning, etc.

Mathematical Model. The assumptions which make up the core of the model are as follows:
1. A slab geometry of the polymer matrix with center line at x=0 and outer surface at x=a. See Figure 1.
2. The drug and other molecules are assumed to be in a single phase with the polymeric structure.
3. No volume change of the system as water diffuses in. Poly(ortho ester)s are rather hydrophobic (3).
4. Perfect sink conditions externally with finite mass transfer boundary layers.
5. Transport in the polymeric system is assumed to be by Fickian diffusion, although the diffusivity of the various species depends on the extent of hydrolysis of the polymeric linkages.
6. The chemical reactions are assumed to be elementary and the overall reaction scheme is given in Figure 1.

The mathematical model then is a set of coupled, nonlinear, one dimensional, unsteady-state mass balances of the form

$$\frac{\partial C_i}{\partial t} = \frac{\partial}{\partial x} D_i(x,t) \frac{\partial C_i}{\partial x} + v_i \qquad (1)$$

where C_i is the concentration of the ith species, D_i is the diffusivity of the ith species, x is the position, t is the time, and v_i is the net sum of the rate of production and degradation of the ith species given by the reaction scheme as shown in Figure 1. Since the diffusivity is dependent on the extent of hydrolysis, the diffusivity is assumed to take the form (1)

$$D_i^\circ = D_i^\circ \exp \left[\frac{\mu(C_D^\circ - C_D)}{C_D^\circ} \right], \qquad i = A, B, C, E \qquad (2)$$

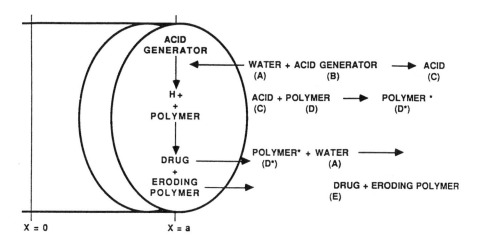

Figure 1. Scheme depicting erosion process of polymer disc containing acid labile linkages. (Reproduced with permission from ref. 7. Copyright 1991 Elsevier Science Publishers.)

where D_i is the diffusivity of the ith species, D_i^0 is the diffusivity through unreacted polymer, and μ is a proportionality constant.

The boundary conditions for the system are then:

1.   Symmetry at the centerline:

$$D_i(0,t)\frac{\partial C_i}{\partial x}(0,t) = 0, \qquad 0 < t, \ i = A, B, C, E \qquad (3)$$

2.   Finite mass transfer into the bulk environment which approximates sink conditions:

$$D_i(0,t)\frac{\partial C_i}{\partial x}(a,t) = k_i\left(C_{i,bulk} - C_i(a,t)\right), \qquad 0 < t, \ i = B, C, E \qquad (4)$$

where $C_{i,bulk}$ is the concentration of ith species at time zero in the bulk of the aqueous phase.

3.   For water, there is local equilibrium between the water in the disk and in the environment at the surface.

$$C_A(a,t) = K(C_D)C_A^0$$

where C_A^0 is the concentration of water in the aqueous phase and K is the partition coefficient.

The initial conditions are:

$$C_i(x,0) = 0 \quad for \ 0 < x < a, \ i = A, C, D^*$$
$$C_i(x,0) = C_i^0 \quad for \ 0 < x < a, \ i = B, D, E$$

where C_i^0 is the initial concentration of species i in the matrix.

The above model can calculate the concentration of ortho ester linkages as a function of time and position. Nguyen et al. (5) have shown that the rate of reaction of ortho ester linkages is independent of the copolymers incorporated. As such, it can then be assumed that the polymer reaction is independent of the extent of reaction and therefore molecular weight. As a result, random chain degradation kinetics can be used to calculate the molecular weight at any position and time. If p is the probability of finding a reacted group and d the probability of finding an unreacted group, then the probability of finding a molecule x unit long is given by $p^{(x-1)}(1-p)$ or $(1-d)^{(x-1)}d$. Thus the probability of finding a chain x unit long, N_x, in a polymer solution containing N chains of different lengths is

$$N_x = N\ (1-d)^{(x-1)}d$$

The number average degree of polymerization is then

$$x_n = \Sigma \, x \, N_x \, / \, \Sigma \, N_x$$

and the number average molecular weight is given by

$$M_n = \Sigma \, M_x N_x \, / \, \Sigma \, N_x$$

By similar analysis, weight average molecular weight and the polydispersity in the polymer matrix can be calculated. By summing these results at every point in time and position in the disk, the distribution of molecular weight is calculated. The resulting spatial distribution is then integrated over the entire disk to calculate a weight-average molecular weight for the disk. The predictions generated using simple random scission kinetics provide a means for system performance assessment and allow the identification of the degradation mechanism of polymer matrix, namely, surface erosion versus bulk erosion.

In order to characterize the erosion properties of specific systems, the Thiele modulus ϕ has been used to describe those systems where chemical reaction and transport are both important. By making the above Equations (1 - 4) dimensionless, several Thiele moduli are noted. The one which describes the transport of water compared to its consumption by chemical reaction is

$$\phi^2 = \frac{a^2/D_{Ao}}{1/k_1 C_{Bo}} \tag{5}$$

Equation (5) can also be expressed as

$$\phi^2 = \frac{\tau_{\text{diffusion}}}{\tau_{\text{reaction}}} \tag{6}$$

where $\tau_{\text{diffusion}}$ is the characteristic time constant for diffusion and τ_{reaction} the characteristic time constant for chemical reaction. The ratio is then a direct measure of the rate limiting step for the overall process: if the modulus is much greater than unity, diffusion is the rate limiting step while those less than unity denote a reaction limited system.

Results and Discussion

The above model was solved numerically by writing finite difference approximations for each term. The equations were decoupled by writing the reaction terms on the previous time steps where the concentrations are known. Similarly the equations were linearized by writing the diffusivities on the previous time step also. The model was solved numerically using a linear matrix inversion routine, updating the solution matrix between iterations to include the proper concentration dependent diffusivities and reactions.

The system modelled was described by Nguyen et al. (2). The parameters for the model are taken from independently conducted experiments (2-6). Computer simulations were conducted with the model to mimic the conditions of the experimental situation. The simulation results are presented for a disc device whose dimensions are given in Table I. The molecular weight of each monomeric unit of the polymer is about 350 with two sites of hydrolysis. The concentration of device components and the dimensionless property ratios corresponding to each disc are given in Table II.

For poly (ortho ester)s, the amount of catalyst present can drastically alter the erosion characteristics. As a result, consider the results of the mathematical model at two different anhydride levels, 0.25% and 1.0%.

Erosion of the Device with 0.25% Anhydride. Figure 2 demonstrates the erosion characteristics of a device loaded with a relatively low concentration of catalyst. The experimental results and the model both demonstrate drug release over a one hundred hour time frame. The rate of anhydride depletion is exponential, not linear, as would be expected in surface erosion. Thus it can be surmised that water is permeating into the device faster than it can be consumed by the catalyzed reaction. An examination of the Thiele modulus confirms this hypothesis. The Thiele modulus for this system is about 0.76; the time constant for transport is about 45 hours. After two time constants, the system is 90% saturated with water. Therefore, bulk degradation can be expected. The decrease in molecular weight shown in Figure 3 is also well modelled. The initial linear decrease corresponds to the build up of acid in the system and random chain scissioning near the saturated boundary. However, as the matrix is saturated, degradation occurs by a slow but bulk destroying process.

Erosion of the Device with 1.0% Anhydride: As shown in Figures 4 and 5 the model results agree favorably with those found experimentally. In this case, as shown in Figure 4, the anhydride content drops off linearly with time as would be expected if the water diffusion rate were controlling the overall process. The Thiele modulus for this system is 1.5. Therefore, as water intrudes into the matrix, there is sufficient anhydride present to react with the water to effectively consume it. Therefore the process is diffusion limited as denoted by the Thiele modulus greater than unity. Finally the model also calculates reasonably well the molecular weight changes seen experimentally, over a time course considerably shorter than the 0.25% case. Since the molecular weight changes in both the model and experimental release coincide with the drug release rate, it can be seen that the drug release is primarily due to the mechanisms included in the model.

Conclusion

A mathematical model has been developed which allows the calculation of the degradation of polymeric drug delivery systems. The model has been shown to accurately simulate both the drug release and molecular weight changes in such systems. The concentration of anhydride levels affect the erosion characteristics of

Table I: Device Dimensions and Properties

Diameter of Disc	1 cm
Average thickness	0.14 cm
Device volume	0.1099 cm^3
Device weight	57 mg
Dye loading	0.5%
Device Composition	refer to Table II

Table II. Device Composition and Dimensionless ratios

	Device 1	*Device 2*
Phthalic Anhydride	0.25%	1.0%
C_B	8.75 x 10^{-6} moles/cm^3	3.50 x 10^{-5} moles/cm^3
C_D	2.94 x 10^{-3} moles/cm^3	2.91 x 10^{-3} moles/cm^3
C_{Ao}	0.0555 moles/cm^3	0.0555 moles/cm^3
$\beta = C_B/C_D$	0.00297	0.01199
$\alpha = C_A/C_B$	6339.59	1585.70
ϕ_1	0.756	1.512
ϕ_2	7.592	7.563
ϕ_3	75.92	75.63

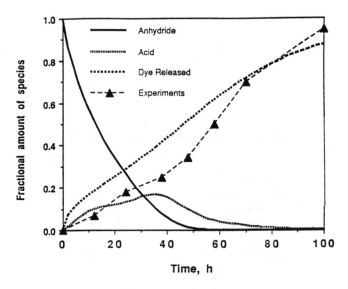

Figure 2. Cumulative release of active species of the disc containing 0.25% anhydride. (Reproduced with permission from ref. 7. Copyright 1991 Elsevier Science Publishers.)

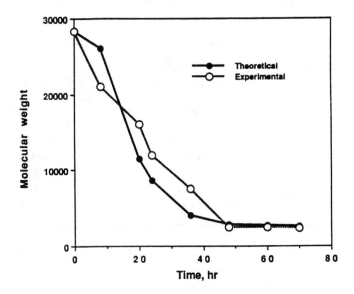

Figure 3. Molecular weight of the degrading polymer disc containing 0.25% anydride as a function of time. (Reproduced with permission from ref. 7. Copyright 1991 Elsevier Science Publishers.)

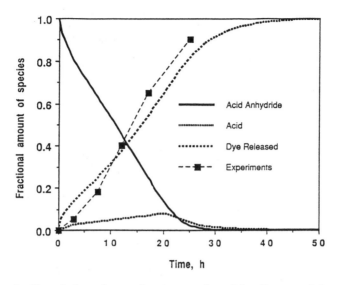

Figure 4. Cumulative release of active species of the disc containing 1.0% anhydride. (Reproduced with permission from ref. 7. Copyright 1991 Elsevier Science Publishers.)

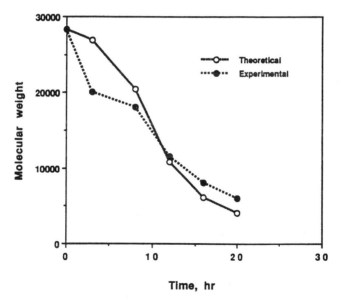

Figure 5. Molecular weight of the degrading polymer disc containing 1.0% anhydride as a function of time. (Reproduced with permission from ref. 7. Copyright 1991 Elsevier Science Publishers.)

the device in two ways: (1) it reduces the life span of the device, and (2) the governing release mechanism is surface erosion for higher anhydride concentration and bulk for lower anhydride concentration. Further, the Thiele modulus, a standard dimensionless group often used in systems which include both reaction and transport, can be used to distinguish between surface and bulk eroding systems when the rate-limiting processes can be determined.

Literature Cited

1. Thombre, A. G.; Himmelstein, K. J. *AIChE Journal* **1985**, *31(5),* 759.
2. Nguyen, T. H.; Higuchi, T.; Himmelstein, K. J. *J. Controlled Release* **1987**, *5,* 1.
3. Nguyen, T. H.; Himmelstein, K. J.; Higuchi, T. *Int. J Pharm.* **1985**, *25,* 1.
4. Sparer, R. V.; Shih, C.; Ringeisen, C. D.; Himmelstein, K. J. *J. Controlled Release* **1984**, *1,* 23.
5. Nguyen, T. H.; Shih, C.; Himmelstein, K. J.; Higuchi, T. *J. Pharm. Sci.* **1984**, *73,* 1563.
6. Nguyen, T. H.; Himmelstein, K. J.; Higuchi, T. *J. Controlled Release* **1986**, *4,* 9.
7. Joshi, A.; Himmelstein, K. J. *J. Controlled Release*, in press.

RECEIVED March 19, 1991

Chapter 17

Bicontinuous Controlled-Release Matrices Composed of Poly(D,L-lactic acid) Blended with Ethylene–Vinyl Acetate Copolymer

Harli M. Dollinger and Samuel P. Sawan

Polymer Science and Plastics Engineering Program, Department of Chemistry, University of Lowell, Lowell, MA 01854

A series of poly(D,L-lactic acid)(PLA) - ethylene/vinyl acetate copolymer blends, from 70% to 30% PLA, and the homopolymers have been examined as possible matrices for controlled release of insoluble biologically active agents. Various blending techniques, such as extrusion and solution blending, were used to create the samples. The hydrolytic degradation of PLA and concurrent release of lactic acid from these blends were studied in both phosphate buffer solution, and buffer solution containing a phase transfer catalyst. Acetone extraction studies, mass loss profiles, differential scanning calorimetry, and scanning electron microscopy have given both insight into the effect of blending technique on morphology, and the concurrent effect of morphology on degradation. Solution blended samples have a much finer porous structure than do the extruded samples, and therefore allow for a more controlled and longer lived lactic acid release. With all blending techniques, through the entire range of blends, both phases show continuity.

The use of polymers for biomedical applications has been widely accepted since the 1960's (1), and specifically for controlled release applications since the 1970's (2). The primary goal of this research was to create a controlled release matrix from polymers with pre-existing Food and Drug Administration (FDA) histories, which would be capable of releasing insoluble active agents, and upon exhaustion of the device, leave a stable, inert, removable skeleton. The application of such a matrix would be as an intracervical device which would prevent both conception and ascending infection.

Historically, natural polymers such as carboxymethylcellulose, starch and proteins, synthetic polymers such as polyvinyl alcohol, polyamides and many

0097–6156/91/0469–0181$06.00/0

polyesters and synthetic elastomers such as polybutadiene, silicone rubber and acrylonitrile have been used in controlled release devices as coatings, microencapsulations, films in laminated structures and as erodible devices in both agricultural and biomedical applications (3). More recently new classes of polymers have been studied, such as poly(anhydrides) and poly(orthoesters) (4,5), poly(iminocarbonates) (6), and hydroxybutyrate-hydroxyvalerate homo- and copolymers (7,8).

To review, research in our laboratories began in the early 1980's with the hydrolytically unstable PLA as a matrix for releasing insoluble agents (9,10). PLA has a long history of use as surgical repair materials and for fertility and cancer control (11,12,13). Again, it was desirable to use compatible materials with preexisting FDA histories, which could release insoluble, nonsystemic agents, and release only nontoxic by-products. PLA loses its mechanical integrity during bulk erosion after approximately 4 to 5 months, making it unsuitable for long-term controlled release.

Ethylene/vinyl acetate copolymer (EVAc) has gained considerable attention in this field due to its unique ability to release soluble macromolecules at controlled rates for prolonged periods (14,15,16). The drug is typically solution blended with the EVAc and upon introduction to an aqueous environment, water diffuses in through the hydrophilic drug phase, and dissolves it. The drug is released and a series of interconnecting pores remain through which more water can diffuse. The release rate and pore size have been shown to be dependent on the drug particle size and loading (17).

PLA and EVAc were blended by various techniques believing that the same principles would apply, that bicontinuous matrices would be generated where the hydrolytically unstable PLA would degrade and an insoluble agent blended with it would be released along with the PLA degradation products, leaving behind a stable EVAc skeleton. Preliminary results of early hydrolysis experiments have been reported previously (18). This report gives an overview of the experimentation completed on this system, but concentrates on the uniqueness of the PLA/EVAc blends. The effects of thermal processing on PLA molecular weight degradation, acetone extraction results, differential scanning calorimetry results of the initial blends, and the results of a 48 week hydrolysis study are given. The type of blending technique is shown to determine the pore sizes generated, the mass loss profile and the amount of water held within the system.

Experimental

PLA (courtesy of Boehringer Ingleheim Kg.) and EVAc (courtesy of DuPont) were blended by four different techniques elucidated below:
(A) Solution blending in 15% methylene chloride followed by precipitation in water with stirring. The cocoons formed were vacuum dried and passed through a melt indexer at 90°C to form cylinders with 3-4 mm diameters. They were then cut into lengths of 20 mm using a single edged razor.

(B) Simple extrusion at 110°C using a CSI Minimax extruder, creating samples the same size as above.

(C) Solution blending followed by extrusion.

(D) Double extrusion followed by melt index.

The compositions studied included 100/0, 70/30, 50/50, 30/70, and 0/100 (wt% PLA/EVAc).

Gel permeation chromatography was performed in tetrahydrofuran using a Waters pump system and a Model 410 differential refractive index detector for the eluant. Five Ultrastyragel columns with nominal porosities ranging from 500 to 10^5 angstroms were used for all the samples and the polystyrene standards.

Solvent extraction experiments were performed on all processed homopolymers and blends. The samples were first weighed then immersed in 20 mL acetone for 24 hours at room temperature. The remaining solid rod was removed from the acetone, vacuum dried and reweighed. The remaining experimentation was performed only on samples from (A) and (D) above.

Scanning electron microscopy was performed on base hydrolyzed (NaOH) samples in order to study the structure after PLA degradation and release using a Hitatchi S570 SEM.

Thermal characterization was carried out using a Perkin Elmer DSC-2 Differential Scanning Calorimeter. Three heating runs and three cooling runs at 20°K/min were done in duplicate on the full range of samples from 230°K to 490°K. All experiments were run under nitrogen in a dry box. The crystalline melting point (T_m of EVAc) and the glass transition temperature (T_g of PLA) were determined from the second cooling runs. Cp and ΔHf were also determined in this manner. The sample sizes ranged from 2.8 to 9.6 mg. Aluminum sample pans were used.

Eight samples from each set (A and D) were split into two groups, one incubated in 0.15 M phosphate buffered solution (PBS) and the other in PBS containing 0.1% of the phase transfer catalyst, tricapryl methyl ammonium chloride (Aldrich Aliquot 336) (PBS, catalyzed), both at pH 7.4. The samples were suspended in centrifuge tubes with 10 mL of incubation solution by nylon string and the tubes were placed in a shaker bath at a constant temperature of 37°C. Periodically, the samples were taken out of the tubes, placed on lint free paper towels, allowed to dry for exactly one hour, weighed and placed in fresh incubation media. The total time of the study was 48 weeks (330 days).

At the end of the study, the samples were air dried for 24 hours, dried under vacuum for 24 hours and weighed. The procedure was repeated until a constant weight was reached.

Results and Discussion

PLA molecular weights were evaluated as a function of thermal processing treatment since molecular weight is of primary importance for degradation lifetimes. As can be seen in Table I, there are significant molecular weight changes as a function of the process history for PLA. All thermal processing techniques employed resulted in a degradation of PLA molecular weight. Sample D, which was thermally processed three times, showed the largest molecular weight reduction. The molecular weight of PLA may significantly affect the hydrolytic lifetime of the material and this must be considered in the preparation of polyblends containing PLA.

Acetone extractions were performed on polyblends of PLA with EVAc to quickly assess the nature of the blend and the total amount of extractable PLA in such blends. Acetone is a good solvent for PLA but is not a solvent for EVAc, although EVAc does swell in this solvent. Table II shows the normalized amount of extractable PLA for various blend compositions and blending techniques. In most cases, the PLA cannot be totally recovered from these blends. However, the percent recovery is surprisingly high in most cases, especially where the PLA was the minor component in the polyblend. It was also interesting to note that even when the PLA was the major component, not all of the PLA was recoverable with acetone extractions. Also note that the molecular weight of the PLA affected the amount extractable. As seen for the extrusion blending technique, two different molecular weights of PLA were evaluated. The higher molecular weight material showed approximately 40% less extractables as compared to the lower molecular weight sample. EVAc consistently showed between 2 to 5% acetone extractables.

Comparison of the blending techniques also reveals some interesting insight into the behavior of these blends. The most homogeneous blends, i.e. those prepared by solution mixing of the polymers before any thermal processing, allowed for less extraction of the PLA. Thus, as might be expected, PLA was becoming trapped within EVAc rich domains.

These data suggested that such blends may show continuity of both phases over a wide range of compositions. The physical appearance of the blends before and after acetone extraction also supported the idea that both polymer phases could be continuous. The blended materials were tough, rubbery rods with physical properties dependent upon the blend composition. Acetone extracted rods were spongy, even when high, e.g. 70% by weight, of PLA was used in the blend. These studies give only qualitative, not true quantitative results since acetone causes the system to swell. The most important information gained is the fact that stable skeletons remained throughout the entire range of blends after PLA extraction.

Differential scanning calorimetry is useful in the study of blend morphology. Typically, a DSC curve rises linearly at low temperatures and then rises more steeply at the glass transition, and as crystallization takes place, a sharp drop in the specific heat curve occurs. PLA is an amorphous polymer which typically shows a T_g midpoint about 45 to 50°C. EVAc shows a distinct crystalline melting point around 24-25°C. The degree of miscibility of two polymers can be determined by the shift in T_g or by the shift or even disappearance of T_m. From Table III and Figures 1a and 1b, it can be seen that over the entire range of blends, for both solution blended and extruded samples, no change in either T_g of the PLA or T_m of the EVAc occurs. This shows that PLA/EVAc blends are completely immiscible. This is a primary qualification of bicontinuous blends.

The DSC results do not give information regarding the relative sizes of the phases or the physical blend structure. This information can only be inferred from the mass loss profiles and the amount of water retained in the pores and interconnecting

Table I. Molecular Weight Change in PLA as a Function
of Thermal Processing[1]

Sample	Mn	Mw	Mz	Mw/Mn
PLA Control	83,700	120,000	162,700	1.43
A	67,800	117,500	174,200	1.73
B	61,300	100,000	143,000	1.63
C	60,700	97,800	138,400	1.61
D	52,200	76,600	103,000	1.47

[1] By GPC in THF

Table II. Acetone Extractable Percentage of
PLA in PLA/EVAc Polyblends

Sample	MW	---------Blend Composition - (PLA/EVAc)----------				
		100/0	70/30	50/50	30/70	0/100[1]
A	117,500	100	86.5	88.6	83.3	1.8
B	100,000	100	100	---	97.6	5.3
	210,000	100	60.3	---	59.6	3.7
C	97,800	100	57.2	61.2	19.5	4.2
D	76,600	100	100	78.6	87.5	2.7

[1] Amount of extractable residue from EVAc

**Table III. DSC Analytical Results for the Second Cooling Runs
for Solution and Extrusion Blended PLA/EVAc Samples**

Sample PLA/EVAc	Tg Onset °C	Tg Midpoint °C	Cp Cal/g °C	Tm Onset °C	Tm Midpoint °C	ΔHf Cal/g
Solution Blended						
0/100	=====	=====	=====	29.28	24.97	1.94
30/70	51.66	46.71	-2.49e-2	29.16	24.82	1.51
50/50	51.29	46.83	-4.18e-2	29.03	24.85	1.12
70/30	50.71	46.67	-6.28e-2	29.08	24.68	0.54
100/0	49.75	45.86	-6.77e-2	=====	=====	=====
Extruded						
0/100	=====	=====	=====	29.76	24.38	1.96
30/70	50.57	46.62	-2.97e-2	29.62	24.17	1.31
50/50	49.61	45.31	-4.39e-2	29.41	24.49	1.12
70/30	51.00	45.96	-7.27e-2	29.29	24.65	0.61
100/0	49.77	45.35	-8.23e-2	=====	=====	=====

tunnels of the system. Figures 2 through 5 show the normalized apparent mass loss profiles for solution blended and extruded samples during 330 days of incubation, with and without the phase transfer catalyst. Table IV shows the real and apparent mass loss, the amount of water in the systems and the total percent of normalized PLA lost over 48 weeks of incubation.

The phase transfer catalyst was used in both in order to create a better in vitro assay and to determine if such a catalyst could decrease the total time of these studies without altering either the temperature or pH of the incubation media. As reported previously (18), the catalyst appears to effect only the initial induction period, however as degradation proceeds, the kinetics of the bulk erosion become the overriding parameter. There appears to be a slight differentiation between the type of blend or the blend stoichiometry in regards to the effect of the phase transfer catalyst. The catalyst seems to increase the initial degradation to a greater extent in the more hydrophobic blends, those with a greater amount of EVAc.

From the mass loss profiles and the total amount of water retained in the system it becomes apparent that the two types of blends are in fact quite different. The solution blended samples show a longer induction period (the amount of time it takes prior to mass loss), a more constant mass loss and a greater amount of water retained in the system. From this it can be assumed that the blend has a much finer, more tortuous structure. Mass loss from PLA hydrolysis begins only after the polymer has degraded to low molecular weight fragments of either monomer or dimer. It is hypothesized that it will take a greater amount of time for the PLA to degrade and the oligomers to diffuse through a more tortuous structure. By the same token, a matrix with a greater amount of surface area, and hence smaller pores, (after PLA degradation and subsequent lactic acid release) will retain a greater amount of water. The solution blended samples show this phenomena. The extruded samples have shorter induction periods, a less constant mass loss profile and less water retention, which indicates a blend structure with much larger pores.

It can be deduced that the molecular weight of the PLA, within the range studied, does not effect the rate of degradation and subsequent mass loss. Figures 2 through 5 show that the PLA homopolymer from set A), solution blended, actually degrades at a faster rate than does the extruded samples, set D). This is probably due to the lower amount of overall compression during processing, and hence a less tightly packed system.

The micrographs, Figures 6 and 7, show a side view of a 50/50 extruded blend, and a 70/30 blend after base hydrolysis. The porous, tunneling structure can clearly be seen.

Conclusions

This research has given insight to a very unique bicontinuous blend system. It has also shown that the pore structure and release rates from EVAc matrices are not only dependent on particle size and loading but also on the blending technique. It becomes apparent that it is possible to tailor these systems not only for release rates but also for initial time of release of insoluble biologically active agents.

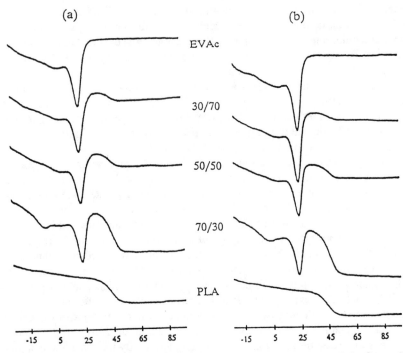

Figures 1a and 1b. Differential Scanning Calorimetry Results of the Second Cooling Runs for (a) Solution Blended and (b) Extrusion Blended PLA/EVAc Homopolymers and Copolymers.

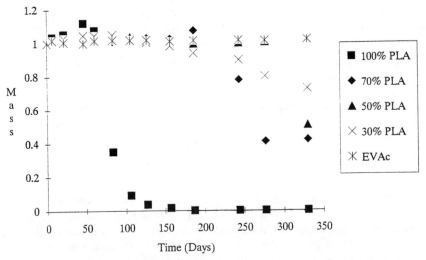

Figure 2. Normalized Mass of Solution Blended / Uncatalyzed Samples During 330 Days of Incubation.

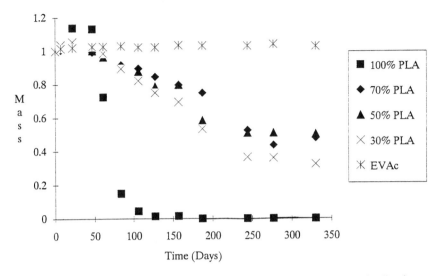

Figure 3. Normalized Mass of Solution Blended / Catalyzed Samples During 330 Days of Incubation.

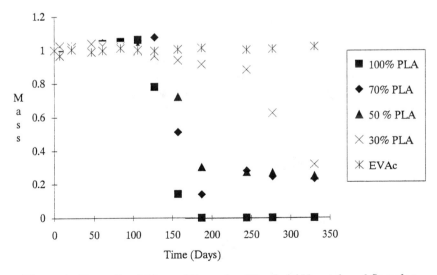

Figure 4. Normalized Mass of Extrusion Blended / Uncatalyzed Samples During 330 Days of Incubation.

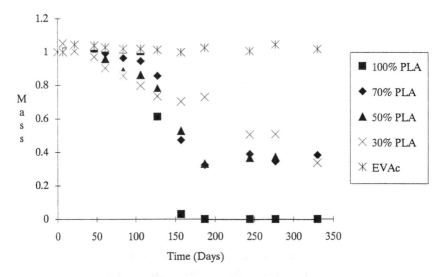

Figure 5. Normalized Mass of Extrusion Blended / Catalyzed Samples During 330 Days of Incubation.

Figure 6. Scanning Electron Micrograph of a 50/50 (PLA/EVAc) Blend After Base Catalyzed Hydrolysis. (Side View)

**Table IV. Final Apparent Percent Mass Loss, Percent Water
Retained in Pores and Actual Percent PLA Loss of
All Samples Studied After Incubation for 330 Days[1]**

Sample (PLA/EVAc)	% Apparent Mass Loss	% Real Mass Loss	% Water in Pores	Actual % PLA Loss
Solution Blended				
(uncatalyzed)				
70/30	41	67	27	96
50/50	25	46	21	91
30/70	20	28	8	92
(catalyzed)				
70/30	38	67	31	93
50/50	24	44	20	89
30/70	20	25	5	85
Extruded				
(uncatalyzed)				
70/30	54	68	14	97
50/50	38	45	7	89
30/70	21	29	8	95
(catalyzed)				
70/30	43	65	22	93
50/50	32	42	10	85
30/70	20	28	8	92

[1]Average of four samples, rounded to nearest percent.

Figure 7. Scanning Electron Micrograph of a 70/30 (PLA/EVAc) Blend After Base Catalyzed Hydrolysis.

Acknowledgments

We gratefully acknowledge the financial support provided by the Population Council, the Plastics Institute of America Inc., and the University of Lowell Research Fellowship Programs.

Literature Cited

1. Paul, D.R. In *Polymers in Controlled Release Technology;* Paul and Harris Eds., Controlled Release Polymeric Formulations, Symposium Series 33; American Chemical Society: Washington, D.C., 1976, pp 1-12.
2. Langer, R.S.; Peppas, N.A. *Biomat.* **1981**, *2,* 201.
3. Zeoli, L.; Kydonieus, A. In *Physical Methods of Controlled Release;* Das, K., Ed.; Controlled Release Technology, Bioengineering Aspects; John Wiley & Sons: New York, 1983; pp 61-121.
4. Rosen, H.; Chang, J.; Wnek, G.; Linhardt, R.; Langer, R. *Biomat.* **1983**, *4,* 131.
5. Domb, A.; Langer, R. *Macromol.* **1989**, *22,* 2117.
6. Li, C.; Kohn, J. *Macromol.* **1989**, *22,* (5), 2029.
7. Holland, S.; Jolly, A.; Yasin, M.; Tighe, B. *Biomat.* **1987**, *8,* (4), 289.
8. Dave, P.; Ashar, N.; Gross, R.; McCarthy, S. *Poly. Prepr.* **1990**, *31,* (1), 442.
9. Barry, J. J., Ph.D Thesis, *Evaluation and Characterization of the Degradation and Silver Release From a Poly (d,l- Lactic Acid) - Silver Matrix,* Department of Chemistry, University of Lowell, 1988, Bl-1, 92.
10. Sawan, S. P.; Barry, J. J. *Poly. Prepr.,* **1988**, *29,* (1), 299.
11. Wise, D.; Fellmann, T.; Sanderson, J.; Wentwotrth, R. In *Lactic/Glycolic Acid Polymers,* Drug Carriers in Biology and Medicine; Academic Press: London, 1976; pp 237-270.
12. Kukarni, R. K.; Moore, E.; Hegyeli, A.; Leonard, F. *J.Biomed. Mat. Res.* **1971**, *5,* 169.
13. Van der Meer, R.; Aarts, A. M.; German, A. L., *J. Poly. Sci., Poly. Chem. Ed.* **1980**, *18,* 1347.
14. Wheeler, R. G.; Friel, P. G., *Release of Drugs From IUD's Using an Ethylene / Vinyl Acetate Matrix,* International Fertility Research Program, Research Triangle Park, NC, 1982.
15. Langer, R. *Chemtech* **1982**, *12,* 98.
16. Hsu, T. P.; Langer, R. *J. Biomed. Mat. Res.,* **1985**, *19,* 445.
17. Rhine, W.; Hsieh, D.; Langer, R. *J. Pharm. Sci.,* **1980**, *69,* 265.
18. Dollinger, H. M.; Sawan, S. P. *Poly. Prepr.* **1990**, *31,* (1), 429.

RECEIVED March 19, 1991

Chapter 18

Amphiphilic Networks

Synthesis and Characterization of and Drug Release from Poly(N,N-dimethylacrylamide)-1-polyisobutylene

Béla Iván[1], Joseph P. Kennedy[2], and Paul W. Mackey

Institute of Polymer Science, University of Akron, Akron, OH 44325–3909

Amphiphilic networks comprising poly(N,N-dimethylacrylamide) (PDMAAm) linked (_1_) by polyisobutylene (PIB) segments have been synthesized by the radical copolymerization of DMAAm and narrow molecular weight ($M_w/M_n \sim 1.1$) distribution methacrylate-telechelic PIB macromonomers (MA-PIB-MA). Conditions have been optimized for the preparation of various PDMAAm-1-PIB networks of good mechanical properties. DSC traces exhibit a high and a low temperature T_g (i.e., from 90 to 115 °C and from -52 to -63°C, respectively) which indicates microphase separation into PDMAAm and PIB domains. The amphiphilic and co-continuous nature of PDMAAm-1-PIB networks was demonstrated by swelling both in water and n-heptane. Drug delivery devices were prepared by loading theophylline into tough flexible PDMAAm-1-PIB networks and studying the release of this material into water. Release studies indicate sustained drug release and kinetics controlled by combination of the M_n of the PIB segments and the PIB content.

New drug delivery systems are of great scientific and commercial interest. Amphiphilic networks composed of about 50/50 hydrophobic PIB and hydrophilic poly(2-(-dimethylamino)ethyl methacrylate) (DMAEMA) polymer segments were found to be biocompatible and to a large extent avascular (_1_). These PIB-1-PDMAEMA networks (_1_, in line with propositions of Weber and Stadler (_2_), and Sperling (_3_), denotes PDMAEMA chains linked by PIB chains) gave pH dependent

---

NOTE: This chapter is Part III in a series.

[1]Current address: Central Research Institute for Chemistry of the Hungarian Academy of Sciences, H–1525 Budapest, P.O. Box 17, Hungary

[2]Address correspondence to this author.

controlled release of bromophenol blue and folic acid over a period of several days
(*4*). We speculated that similar amphiphilic networks containing the hydrophobic and
hydrophilic polymer segments in a balanced ratio may also be biocompatible and
therefore of interest for biomedical applications. A thorough search into the
literature, however, indicated that there is a dearth of information about such
materials most likely due to the synthetic difficulties of linking highly incompatible
hydrophilic and hydrophobic polymers (*2,4-6*).

This paper concerns the synthesis and characterization of a series of poly(*N,N*-
dimethylacrylamide)-1-polyisobutylene (PDMAAm-1-PIB) amphiphilic networks.
The use of these networks as potential drug delivery systems is also demonstrated.

Experimental

Materials. Tetrahydrofuran (THF) was freshly distilled from sodium, and
triethylamine from P_2O_5. Methacryloyl chloride (Aldrich Chemical Co.) and *N,N*-
dimethylacrylamide (DMAAm), (Aldrich Chemical Co.) were distilled under reduced
pressure prior to use. 2,2'-Azobisisobutyronitrile (AIBN) (Eastman Kodak) and
theophylline (Mp 275 °C) (Aldrich Chemical Co.) were recrystallized from methanol
and water, respectively. Hexanes, *n*-heptane, and ethanol were used as received.

Synthesis of PIB prepolymers. *tert*-Chlorine-telechelic PIB (M_n=4,000; M_w/M_n
1.09) (*7*), and an allyl-telechelic PIB (M_n=9,500; M_w/M_n 1.14) (*7,8*) were prepared
by living carbocationic polymerizations. The *tert*-chlorine ended PIB was
quantitatively dehydrochlorinated (*9*) to -C(CH$_3$)=CH$_2$ terminated polymer. Both
olefin-telechelic PIBs were then hydroborated and oxidized (*10*) to prepare the
primary hydroxyl termini. The hydroxyl-telechelic polymers were esterified with
methacryloyl chloride to methacrylate-telechelic PIBs, MA-PIB-MA (*11*).

Network synthesis. The networks were prepared by the radical copolymerization of
MA-PIB-MA with DMAAm in THF. THF is a common solvent for both PIB and
DMAAm thereby allowing for a homogeneous reaction mixture. The polymerization
was carried out in disk shaped teflon molds, 1.5 inches in diameter and 0.75 inches in
depth. The molds were filled with the reaction mixture containing the MA-PIB-MA,
DMAAm, and AIBN dissolved in THF, and sealed by a teflon-coated rubber lid. The
mold assembly was placed in a press to prevent evaporation of monomer and solvent
and heated to 60 °C for 72 hours in an oven (DN-43H Scientific Products). The mold
was removed from the oven, allowed to cool while sealed, opened, and the THF was
allowed to evaporate. To remove unreacted MA-PIB-MA, unreacted DMAAm, and
PDMAAm homopolymer the networks were extracted sequentially in a soxhlet
extractor with hexanes for 24 hours and ethanol for 24 hours. The order of the
extraction solvents chosen (first hexanes followed by ethanol) was to insure that the
last solvent could effectively be removed by soaking in water prior to biological
testing. Table I gives the experimental details for the synthesis of two series of
PDMAAm-1-PIBs. The dry networks were transparent tough materials.

Network Characterization. The overall composition of the networks was determined by elemental analysis. Thermal analysis was conducted by DSC (DuPont 1090 Thermal Analyzer) under nitrogen gas at a heating rate of 20 °C/min from -100 °C to 150°C. The swelling behavior of the samples was studied by using *n*-heptane at room temperature and water at 37 °C. The networks were dried in vacuo at 110 °C to constant weight and placed in the solvent. They were weighed periodically until equilibrium swelling was achieved. The swelling curve was obtained by plotting the grams of solvent absorbed per gram of network versus time. The fully swollen networks were optically clear gels that exhibited good mechanical integrity for manual handling.

Release experiments. Networks A-4-56 and A-9.5-49 (3mm x 3mm x 10mm blocks cut out of the original disks) were loaded with theophylline by soaking in theophylline-saturated water solution for three days. Saturation of the loading solution was maintained by a small amount of solid theophylline in the solution. The loaded networks were dried to constant weight in a vacuum oven at 110 °C. The percent loading was determined by measuring the total amount of drug released and dividing it by the mass of the drug released plus the mass of the unloaded network. Each loaded network was placed in a test tube containing 75 mL of distilled water and stirred vigorously. The water was changed periodically to maintain infinite sink conditions and the concentration of the theophylline released determined using a Hewlett Packard 4550A UV/visible spectrophotometer at the absorbance maximum of 274 nm. The release curves were obtained by plotting M_t/M_∞ against time, where M_t is the amount of drug released at a given time and M_∞ is the total amount of drug released.

Results and Discussion

Synthesis. The synthesis of amphiphilic networks involves the homogeneous mixing and subsequent linking of two incompatible polymers. In the present instance the synthesis involved the copolymerization of MA-PIB-MA with DMAAm to yield PDMAAm chains linked by PIB chains. Requirements that must be met for the synthesis of such an architecture are: (1) the monomer and the MA-PIB-MA must copolymerize in a random manner, (2) the kinetic chain must be of sufficient length for the incorporation of at least two MA-PIB-MA units, and (3) phase separation during copolymerization (i.e., linking) must be prevented. The copolymerizability requirement was met by the selection of a monomer, DMAAm, whose polymerizable function is similar to that of the MA-PIB-MA. The sufficient kinetic chain length for the PDMAAm sequence was achieved by adjusting the AIBN concentration according to the relationship $1/DP_n \sim \sqrt{[I]/[M]}$, where DP_n, [I], and (M) are the degree of polymerization, initiator and monomer concentrations, respectively. The AIBN concentration decreases as the overall concentration of polymerizable groups decrease. And finally, phase separation of PDMAAm and PIB was prevented by carrying out the crosslinking reaction in THF, a common solvent for both polymers. The sealed mold was designed to prevent the evaporation of the THF and monomer and to maintain system homogeneity during the crosslinking (that is the linking reaction).

Table I summarizes conditions for the synthesis of two series of PDMAAm-1-PIBs. The samples are identified by a code consisting of a letter and two numbers: the letter (A) denotes the hydrophilic monomer DMAAm, whereas the two numbers indicate the M_n of the starting MA-PIB-MA (divided by 1,000) and the weight percent of PIB in the network (determined by elemental analysis). For example, A-4-26 denotes an amphiphilic network prepared with DMAAm as the hydrophilic moiety, containing a M_n=4,000 MA-PIB-MA, whose composition is 26% PIB.

Table II shows data obtained by sequential *n*-hexane and ethanol extractions. The amount of extractables was below 10% in most cases indicating satisfactory MA-PIB-MA incorporation into the PDMAAm backbone.

Characterization. The overall structure and amphiphilic properties of the networks were characterized by DSC and swelling experiments, respectively. The M_c of the PDMAAm segments (that is the molecular weight between PIB crosslinks of the PDMAAm segment) was calculated by

$$M_c = \left(\frac{W_a \cdot M_n}{2 \cdot W_{PIB}} \right) \qquad (1)$$

where W_a is the weight fraction of PDMAAm, M_n the number average molecular weight of PIB (i.e., M_n = MA-PIB-MA), and W_{PIB} the weight fraction of PIB.

Table II shows T_gs obtained from DSC traces. (Footnotes a and b in Table II show T_gs values of three reference polymers: two PIBs, whose M_ns are similar to the M_ns of MA-PIB-MA used in the network synthesis, and a PDMAAm; the difference in the T_g for the M_n=4,000 and 9,300 PIBs is due to the dependence of T_g on M_n(*12*)). The DSC traces of the networks exhibited two T_gs, one in the range of -63 to -52 °C (PIB domains) and another in the range of 90 to 115 °C (PDMAAm domains) indicating microphase separated structures. The T_gs associated with the PIB phase in the PDMAAm-1-PIB networks were higher than those of the reference homoPIBs which may be due to PIB chain-ends embedded in the glassy PDMAAm phase restricting segmental mobility. The T_g of the PIB phase in the PDMAAm-1-PIB increases by increasing the PIB content which may be due to an increase in crosslink density. In contrast, the T_g for the PDMAAm phase in the network decreases upon increasing the PIB content. Interaction of the (-CH_2-CH-) moiety of the PDMAAm with the flexible PIB and thus the formation of a more flexible structure may explain this phenomenon.

Amphiphilic networks swell both in hydrophilic and hydrophobic solvents. The amphiphilic nature of the PDMAAm-1-PIB networks was substantiated by swelling studies. Figures 1 and 2 show grams of water and *n*-heptane, respectively, absorbed per gram of network (M_n(PIB)=9,500) as a function of time. As the data indicate, the PDMAAm-1-PIB networks swell in both solvents and the degree of swelling changes with network composition. The degree of swelling in water decreases with increasing PIB content, whereas in *n*-heptane it increases with

**Table I. Experimental Conditions for the Synthesis
of PDMAAm-1-PIB Amphiphilic Networks**

Sample	PIB (g)	MA End Group (mol x 10^4)	DMAAm (g)	DMAAm (mol x 10^2)	AIBN (mol x 10^5)
A-4-26	0.6	3.08	1.4	1.41	3.11
A-4-45	0.8	4.10	1.2	1.21	2.13
A-4-56	1.0	5.13	1.0	1.01	1.52
A-4-59	1.2	6.15	0.8	0.81	1.04
A-4-77	1.4	7.18	0.6	0.61	0.73
A-9.5-31	0.6	1.26	1.4	1.41	1.62
A-9.5-40	0.8	1.68	1.2	1.21	1.17
A-9.5-49	1.0	2.11	1.0	1.01	0.84
A-9.5-59	1.2	2.53	0.8	0.81	0.52
A-9.5-76	1.4	2.95	0.6	0.61	0.32

THF solvent; A-4 series total volume 10 mL, A-9.5 series total volume 8 mL; 72 hours, 60°C.

**Table II. Selective Solvent Extraction, Calculated M_c PDMAAm,
and Differential Scanning Calorimetry Data**

Sample	% Extractables Hexane	Ethanol	M_c PDMAAm[c]	PIB Phase[a] T_g °C	PDMAAm[b] Phase T_g °C
A-4-26	5.7	1.8	6,600	-61	104
A-4-45	6.3	3.2	2,400	-54	102
A-4-56	9.3	1.4	1.550	-54	94
A-4-59	8.2	2.1	1,400	-53	90
A-4-77	1.8	0.8	600	-52	ND
A-9.5-31	5.2	18.6	10,600	-63	115
A-9.5-40	4.9	11.9	7,150	-62	110
A-9.5-49	3.1	6.6	4,950	-60	108
A-9.5-59	1.4	5.4	3,300	-59	95
A-9.5-76	1.5	2.5	1,500	-58	ND

[a] T_g of PIB (M_n=9,3000):-65°C, T_g of PIB (M_n=4,000): -69°C.
[b] T_g of Poly(N, N-dimethylacrylamide): 112°C, ND = not determined.
[c] Calculated by using Equation 1.

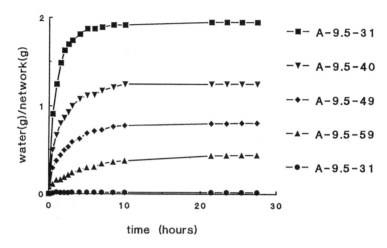

Figure 1. Swelling of PDMAAm-1-PIB amphiphilic networks in water (37°C).

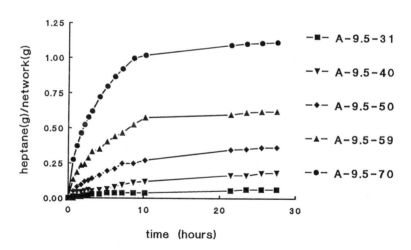

Figure 2. Swelling of PDMAAm-1-PIB amphiphilic networks in *n*-heptane (room temperature).

increasing PIB content. The networks swelled isotropically in either solvent suggesting that the PDMAAm and PIB domains are co-continuous.

The M_c of the PDMAAm segments may be estimated by knowing the overall composition of the networks, the M_n of the MA-PIB-MA, and assuming complete and random incorporation of the MA-PIB-MA in the network. Knowledge of the M_c of the PDMAAm segments and the M_n of the PIB linking segments are necessary to gain insight into the molecular architecture of PDMAAm-1-PIB networks. The architecture of these networks can be controlled by the concentration and the M_n of the MA-PIB-MA employed in the synthesis, which in turn controls the M_c of the PDMAAm segments.

Drug Release from PDMAAm-1-PIB Networks. Amphiphilic networks because of their unique hydrophilic-hydrophobic microphase separated morphology offer many potential biomedical applications. Poly(HEMA-*b*-styrene-*b*-HEMA), poly(HEMA-*b*-dimethylsiloxane-*b*-HEMA, poly(HEMA-*b*-butadiene-*b*-HEMA) triblock copolymers with similar microphase separated morphologies have demonstrated improved antithrombogenic properties over any of the homopolymers by suppressing activation and aggregation of blood platelets at the surface (*17-20*). Because amphiphilic networks share this characteristic of microphase separated hydrophilic and hydrophobic domains it is speculated that they may offer improved biocompatibility over either PIB or PDMAAm. The hydrophilic-hydrophobic structure may also find unique applications as drug delivery devices for both hydrophilic and lipophilic drugs. A drug which is either water or fat soluble can possibly be loaded into the networks and the phase separated structure of the network contribute to the release mechanism. Preliminary release experiments conducted with theophylline as a model for a water soluble drug are presented here. Studies with lipophilic drugs are the subject of ongoing research efforts.

Theophylline, a smooth muscle relaxant, was used as a model for a typical non-ionic water-soluble drug for a study of drug release from our networks. The use of theophylline for this purpose has been documented in the literature (*13,14*). Networks A-4-56 and A-9.5-49, respectively, contained 0.68% and 0.99% theophylline. These low loading levels are due to the low solubility of theophylline in water (0.83% (*15*)). Efforts are underway to increase the extent of loading by increasing the loading time, temperature, and using ethanol as a loading solvent.

Figure 3 shows the amount of theophylline released at time t (M_t) relative to the total amount loaded ($M\infty$) (i.e., $M_t/M\infty$) as a function of time. Evidently theophylline-loaded PDMAAm-1-PIB networks are efficient drug delivery systems, and nearly complete release requires ~24 hours.

The rate and type of release can be analyzed by the expression $M_t/M\infty=kt^n$ (*16*). In the case of pure Fickian diffusion n = 0.5, whereas n > 0.5 indicates anomalous transport, i.e., in addition to diffusion another process (or processes) also occurs. If n = 1 (zero order release), transport is controlled by polymer relaxation ("Case II transport") (*16*). The $\ln(M_t/M\infty)$ versus ln t plots, shown in Figure 4, give n = 0.47 and 0.67 for samples A-9.5-49 and A-4-56, respectively. Evidently theophylline release is controlled by Fickian diffusion in the former network whereas the release is

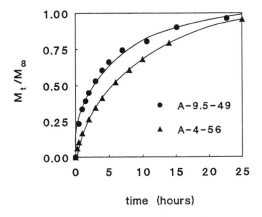

Figure 3. Release of theophylline as a function of time from PDMAAm-1-PIB amphiphilic networks.

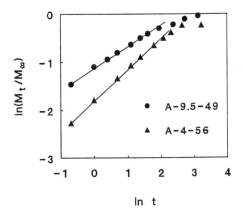

Figure 4. The $\ln(M_t/M_\infty)$ versus ln t for release of theophylline from PDMAAm-1-PIB amphiphilic networks.

partly anomalous in the latter. According to these data, the release of theophylline from PDMAAm-1-PIB networks is controlled by a combination of the M_n of the PIB segments and the PIB content. Further research to determine the effects of PIB content and M_n separately is in progress.

Acknowledgment

Support by NSF Grant DMR-89-20826 is gratefully acknowledged.

Literature Cited

1. Chen, D.; Kennedy, J. P.; Allen, A. J.; Kory, M. M.; Ely, D. L. *J. Biomed. Mat. Res.*, **1989**, *23*, 1327.
2. Weber, M.; Stadler, R. *Polymer,* **1989**, *29*, 1071.
3. Sperling, L. H. In *Multicomponent Polymer Materials;* Paul, D. R.; Sperling, L. H., Ed.; Am. Chem. Soc., Washington, D.C., 1986.
4. Chen, D.; Kennedy, J. P.; Allen, A. J. *J. Macromol. Sci.-Chem.*, **1988**, *A25*, 389.
5. Good, W. R.; Mueller, K. F. In *Controlled Release of Bioactive Materials;* Baker, R., Ed.; Academic Press, 1980, pp. 155-175.
6. Keszler, B.; Kennedy, J. P. *J. Macromol. Sci.-Chem.*, **1984**, *A21*, 319.
7. Kaszas, G.; Puskas, J. E.; Chen, C. C.; Kennedy, J. P. *Polym. Bull.*, **1988**, *20*, 413.
8. Iván, B; Kennedy, J. P. *J. Polym. Sci., Part A: Polym. Chem.*, **1990**, *28*, 89.
9. Kennedy, J. P.; Chang, V. S. C.; Smith, R. A.; Iván, B. *Polym. Bul.*, **1979**, *1*, 575.
10. Iván, B.; Kennedy, J.P.; Chang, V. S. C. *J. Polym. Sci., Polym. Chem. Ed.*, **1980**, *18*, 3177.
11. Kennedy, J. P.; Hiza, M. *Polym. Bull.*, **1983**, *10*, 146.
12. Fox, T. G.; Loshack, S. *J. Polym. Sci.* **1955**, *15*, 371.
13. Korsmeyer, R. W.; Peppas, N. A. *J Controlled Rel.*, **1984**, *1*, 89.
14. Peppas, N. A.; Franson, N. M. *J. Polym. Sci., Polym. Phys.*, **1983**, *21*, 983.
15. *The Merk Index;* Windholz, M.; Budavari, S.; Stroumtsos, L. Y.; Fertig, M. N., Eds.; Ninth Edition; Merk and Co., Inc.: Rathway, N. J.,1976.
16. Peppas, N. A.; Korsmeyer, R. W. In *Hydrogels in Medicine and Pharmacy;* Peppas, N. A., Ed.; CRC Press: Boca Raton, Florida, Vol.III; 1987, pp. 109-135.
17. Okano, T.; Nishiyama, S.; Shinohara, I.; Akaike, T.; Sakurai, Y.; Kataoka, K.; Tsurta, T.; *J. Biomed. Mat. Res.*, **1981**, *15*, 393.
18. Shimada, M.; Miyahara, M.; Tahara, H.; Shinohara, H.; Okano, T.; Kataoka, K.; Sakurai, Y.; *Polym. J.*, **1983**, *15*, 649.
19. Shimada, M.; Sugiyama, N.; Shinohara, I.; Okano, T.; Kataoka, K.; Sakurai, Y. *Eur. Polym. J.*, **1983**, *19*, 929.
20. Okano, T.; Uruno, M.; Sugiyama, N.; Shimada, M.; Shinohara, I.; Kataoka, K.; Sakurai, Y. *J. Biomed. Mat. Res.*, **1986**, *20*, 1035.

RECEIVED March 19, 1991

Chapter 19

Amphiphilic Networks

Synthesis and Characterization of and Drug Release from Poly(2-hydroxyethyl methacrylate)-1-polyisobutylene

Béla Iván[1], Joseph P. Kennedy[2], and Paul W. Mackey

Institute of Polymer Science, University of Akron, Akron, OH 44325–3909

Amphiphilic networks comprising poly(2-hydroxyethyl methacrylate) (PHEMA) linked (1) by polyisobutylene (PIB) segments have been synthesized by radical copolymerization of 2-(trimethylsiloxy)ethyl methacrylate (TMSEMA) with methacrylate-telechelic PIB (MA-PIB-MA) followed by acid catalyzed removal of the trimethylsilyl groups. Conditions have been optimized for the synthesis of two series of PHEMA-1-PIB networks of various compositions. DSC traces exhibit a high and a low temperature T_g (i.e., from 98 to 111 °C and from -61 to -54 °C, respectively) which indicates microphase separation into PHEMA and PIB domains. The amphiphilic and co-continuous nature of PHEMA-1-PIB networks was demonstrated by swelling both in n-heptane and water. The dry networks are transparent tough products; after swelling with water or n-heptane the swollen gels still exhibit satisfactory properties for diverse manipulation. Drug delivery systems were prepared by loading the networks with theophylline. Release studies indicate sustained drug delivery with anomalous diffusion kinetics.

In the previous paper (1) we have described the synthesis, characterization, and certain diffusional characteristics of poly(N,N-methylacrylamide)-1-polyisobutylene amphiphilic networks exhibiting a relatively high degree of swelling in both water and n-heptane. It was of interest to prepare further neutral amphiphilic networks of lower water swelling for sustained drug delivery systems. One candidate for this

NOTE: This chapter is Part IV in a series.
[1]Current address: Central Research Institute for Chemistry of the Hungarian Academy of Sciences, H–1525 Budapest, P.O. Box 17, Hungary.
[2]Corresponding author.

0097–6156/91/0469–0203$06.00/0
© 1991 American Chemical Society

purpose is PHEMA whose utility in hydrogels has amply been demonstrated (2). Copolymerization of HEMA with MA-PIB-MA in tetrahydrofuran (THF), however, leads to massive phase separation due to the large solubility difference between the PIB and PHEMA, and homogeneous networks cannot be obtained.

Phase separation during copolymerization was prevented by reducing the hydrophilicity of HEMA by exchanging the hydroxyl group with a hydrophobic trimethylsilyl group. After copolymerization (i.e., linking) the trimethylsilyl group can easily be removed yielding the desired PHEMA-1-PIB network.

This paper concerns the synthesis and characterization of amphiphilic networks comprising PHEMA and PIB segments. Sustained release studies with theophylline-loaded networks are also described.

Experimental

Materials. 2-Hydroxyethyl methacrylate (HEMA, Aldrich Chemical Co.) and chlorotrimethylsilane (Cl-TMS) (Lancaster Synthesis Inc.) were used as received. The source and purification of the other chemicals has been described (1).

Methacrylate Telechelic Polyisobutylene Synthesis. The synthesis of a 4,000 M_n and a 9,500 M_n MA-PIB-MA has been described (1).

HEMA Silylation and Removal of the - $Si(CH_3)_3$ Group. Trimethylsilylation of HEMA was effected by the dropwise addition of Cl-TMS to HEMA under agitation in the presence of triethylamine as the acid acceptor in THF at 0 °C overnight (3). The triethylamine hydrochloride was removed by filtration and the THF was evaporated by a rotovap. The product was purified by column chromatography and identified as 2-trimethylsiloxyethyl methacrylate (TMSEMA) by ^{1}H NMR spectroscopy (i.e., appearance of the trimethylsilyl -$Si(C\underline{H}_3)_3$ resonance at 0.1 ppm and the disappearance of the hydroxyl -O$\underline{H}$ resonance at 3.3 ppm). The purity of the TMSEMA was determined to be 99% using a Perkin-Elmer 8410 gas chromatograph by the decrease in the retention time of the silylated monomer. The removal of the -$Si(CH_3)_3$ group was carried out by adding a 5 times excess of HCl to a methanol solution of the monomer. The progress of the reaction was followed using GC by the shift to longer retention time for HEMA versus the TMSEMA monomer.

Polymerization of TMSEMA and Desilylation of PTMSEMA. Poly(2-(trimethylsiloxy)ethyl methacrylate) (PTMSEMA) was prepared by the free radical polymerization of TMSEMA in THF ((TMSEMA)=1.5M, (AIBN)=1.125x10^{-4}M, 60 °C, 8 hrs). The homopolymer was precipitated into cold methanol and dried. The removal of the -$Si(CH_3)_3$ group of the homopolymer was carried out at room temperature by dissolving 1 gram of PTMSEMA in a 5% HCl methanol solution using a 5 times excess of HCl. The reaction was run for 4 days with a 25 mL aliquot being withdrawn each day. The aliquot was purified by dialysis (methanol solution against water, 24 hours, 1000 molecular weight cut-off dialysis tubing). Following dialysis the polymer was dissolved in methanol and precipitated into n-pentane. The extent of the reaction was determined using a Gemini 200 ^{1}H NMR spectrophotometer by the appearance of the -O$\underline{H}$ resonance at 3.3 ppm and the disappearance of the -Si(CH$_3$)$_3$ resonance at 0.1 ppm.

Network synthesis. The procedure and equipment used to prepare the networks have been described (*1*). Experimental data and conditions are given in Table I. The abbreviations in the sample column have been explained in the previous paper (*1*). The letter (H) denotes the hydrophilic moiety 2-hydroxyethyl methacrylate. The removal of the -Si(CH$_3$)$_3$ group was accomplished by swelling the networks in a 5% solution of HCl in methanol for two days, followed by swelling in 5% HCl in 2-methoxyethanol for two days. The HCl solution was changed daily. Finally, the networks were soxhlet extracted sequentially with hexanes and ethanol for 24 hrs each to remove all unreacted MA-PIB-MA, monomers, homopolymers, and desilylation byproducts. This order of extraction insures the removal of all extraction solvents by soaking in water prior to biological testing.

Network Characterization. The networks were characterized by DSC (DuPont 1090 Thermal Analyzer) under nitrogen at a heating rate of 20 °C/min. Prior to the experiment the samples were preheated to 130 °C, equilibrated for 10 minutes and cooled to room temperature at 1 °C/min. Swelling experiments were conducted as described (*1*).

Release experiments. The methodology of release experiments has been described (*1*).

Results and Discussion

Synthesis. The procedures used for the preparation of other amphiphilic networks (1) could not be used for the synthesis of PHEMA-1-PIB because of the insolubility of PHEMA in solvents that dissolve PIB e.g., THF. The above described silylation-desilylation procedure was designed to provide mutual solubility of the phases and thus to make the synthesis possible.

Desilylation model studies were carried out on both the silylated monomer and polymer to develop suitable reaction conditions. The desilylation of the TMSEMA was instantaneous as indicated with GC by the increase in the retention time of the monomer. The desilylation the PTMSEMA was equally facile as determined by ^{1}H NMR spectroscopy; Figure 1 shows the disappearance of the -Si(C$\underline{H}_3$)$_3$ resonance at 0.1 ppm and the appearance of the -O$\underline{H}$ resonance at 3.3 ppm without detectible ester hydrolysis even after four days.

Table I summarizes the conditions used for network synthesis. The amount of TMSEMA was determined by the amount of HEMA required for a desired composition. The AIBN concentrations were kept low to insure adequate chain growth during the copolymerization (that is network formation). The obtained networks were transparent, homogeneous, tough, flexible materials demonstrating the utility of the approach. Desilylation of the networks was carried out by the use of

Table I. Experimental Conditions for the Synthesis
of PHEMA-1-PIB Amphiphilic Networks

Sample	PIB (g)	MA End Group (mol x 10⁴)	HEMA (g)	HEMA (mol x 10³)	TMSEMA (g)	AIBN (mol x 10⁵)
H-4-29	0.6	3.08	1.4	10.76	2.18	1.66
H-4-37	0.8	4.10	1.2	9.22	1.87	1.18
H-4-49	1.0	5.13	1.0	7.68	1.55	0.86
H-4-57	1.2	6.15	0.8	6.15	1.24	0.54
H-4-68	1.4	7.18	0.6	4.61	0.93	0.32
H-9.5-38	0.6	1.26	1.4	10.76	2.18	1.62
H-9.5-46	0.8	1.68	1.2	9.22	1.87	1.15
H-9.5-54	1.0	2.11	1.0	7.68	1.55	0.84
H-9.5-64	1.2	2.53	0.8	6.15	1.42	0.53
H-9.5-75	1.4	2.95	0.6	4.16	0.93	0.31

THF solvent; total volume 8 mL, 72 hours, 60°C.

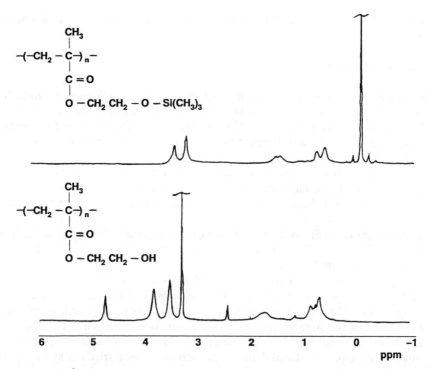

Figure 1. ^{1}H NMR spectra of silylated (TMSEMA) and desilylated (HEMA) homopolymers.

methanol followed by 2-methoxyethanol, because the network swells to a greater extent in this solvent than in methanol. The progress of the reaction could be followed by visually observing the gradual advancement of the swelling-front across the entire width of the sample until its disappearance indicating complete desilylation. Desilylation was substantiated by the large shift in T_g of the silylated phase from 14 °C to the desilylated phase at 106 °C, as shown in Figure 2. For reference, footnotes of Table II give the T_gs of a PTMSEMA (22°C) and a PHEMA (110°C) determined under the same experimental conditions as the networks. The difference between the T_g of the PTMSEMA phase in the silylated network and the homoPTMSEMA may be due to interaction of the PTMSEMA phase in the network by the flexible PIB links resulting in a more flexible structure.

Table II shows data obtained by sequential hexanes and ethanol extractions of the desilylated networks. Attempts to extract the networks prior to desilylation failed because the samples swelled to such a great degree that they lost their mechanical integrity and disintegrated. The amount of hexanes extractable material is less than 3% in all cases which indicates high copolymerization yields. Ethanol extractables were not determined because the presence of the byproducts of the desilylation, which are in the ethanol, would have given artificially high readings.

Characterization. The structure and properties of the networks were investigated by DSC and swelling experiments. The M_c of PHEMA (i.e., the molecular weight of PHEMA sequence between PIB crosslinks) was estimated by

$$M_c = \left(\frac{W_h \cdot M_n}{2 \cdot W_{PIB}} \right) \quad (1)$$

where W_h is the weight fraction of PHEMA, M_n the number average molecular weight of the MA-PIB-MA, and W_{PIB} the weight fraction of PIB, knowing the overall composition of the networks, the M_n of the MA-PIB-MA , and assuming complete and random incorporation of the MA-PIB-MA in the network (*1*).

Table II shows T_g data obtained from DSC traces of the PHEMA-<u>1</u>-PIB networks. The traces showed two T_gs indicating microphase separation into PHEMA and PIB domains. The presence of the PHEMA T_g at ~ 110°C indicates complete desilylation of all networks. The T_gs for the reference PIBs (see footnote a in Table II) are lower than the T_gs of the PIB incorporated into the network. This may be due to the flexible PIB chain-ends embedded in the glassy PHEMA matrix. The increase in the T_g of the PIB phase in the network with increasing % PIB is most likely due to an increase in crosslink density.

The amphiphilic and co-continuous nature of the networks was substantiated by swelling studies. Figures 3 and 4 show the amount of water and *n*-heptane imbibed per gram of network as a function of time for a series of networks (M_n (PIB)=9,500), respectively. As expected, the degree of swelling changes significantly with network composition: the degree of swelling in water decreases from 25% for H-9.5-38 to 1%

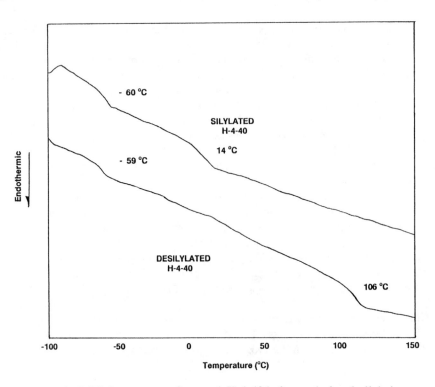

Figure 2. DSC thermogram of network H-4-40 before and after desilylation.

Table II. Selective Solvent Extraction, Calculated M_c PHEMA, and Differential Scanning Calorimetry Data

Sample	%Extractables Hexane	M_c PHEMA[c]	PIB Phase[a] T_g °C	PHEMA[b] Phase T_g °C
H-4-29	0.8	4,850	-61	107
H-4-37	1.2	3,350	-59	106
H-4-49	1.6	2,050	-57	104
H-4-57	2.1	1,500	-55	98
H-4-68	2.1	950	-54	100
H-9.5-38	0.8	7,750	-61	111
H-9.5-46	1.1	5,600	-60	111
H-9.5-54	1.3	4,050	-58	107
H-9.5-64	1.7	2,650	-57	107
H-9.5-75	2.6	1,600	-58	105

[a] T_g of PIB (M_n=9,3000):-65°C, T_g of PIB (M_n=4,000): -69°C.
[b] T_g of Poly(2-hydroxyethyl methacrylate): 110°C,
     T_g of Poly(2-trimethylsiloxyethyl methacrylate): 22°C.
[c] M_c Calculated by using Equation 1.

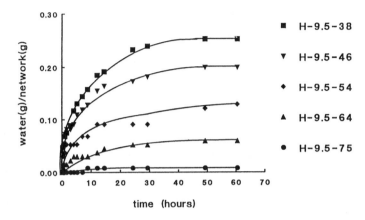

Figure 3. Swelling of HEMA-1-PIB amphiphilic networks in Water (37°C).

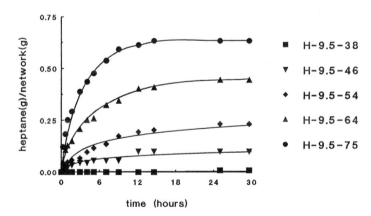

Figure 4. Swelling of HEMA-1-PIB amphiphilic networks in *n*-heptane (room temperature).

for H-9.5-75; conversely, the degree of swelling in heptane is greatest for H-9.5-75 and lowest for H-9.5-38.

The networks swelled isotropically indicating the co-continuous nature of the materials. The range of swelling for the PHEMA-1-PIB networks is significantly less than that of PDMAAm-1-PIB demonstrating that amphiphilic networks exhibiting various desired swelling characteristics can be obtained by the selection of network components.

Table II also shows the calculated M_cs of the PHEMA segments. The M_c of the PHEMA segments and the M_n of PIB determine the molecular architecture of PHEMA-1-PIBs. The architecture of PHEMA-1-PIB networks can be controlled by the concentration and the M_n of the MA-PIB-MA employed in the synthesis which in turn controls the M_c of the PHEMA segments.

Drug Release from PHEMA-1-PIB Networks. Amphiphilic networks due to their distinct microphase separated hydrophobic-hydrophilic domain structure posses potential for biomedical applications. Similar microphase separated materials such as poly(HEMA-b-styrene-b-HEMA), poly(HEMA-b-dimethylsiloxane-b-HEMA), and poly(HEMA-b-butadiene-b-HEMA) triblock copolymers have demonstrated better antithromogenic properties to any of the respective homopolymers (5-8). Amphiphilic networks are speculated to demonstrate better biocompatibility than either PIB or PHEMA because of their hydrophilic-hydrophobic microdomain structure. These unique structures may also be useful as swellable drug delivery matrices for both hydrophilic and lipophilic drugs due to their amphiphilic nature. Preliminary experiments with theophylline as a model for a water soluble drug were conducted to determine the release characteristics of the system. Experiments with lipophilic drugs are the subject of ongoing research.

Theophylline was used as a model drug for release studies. Theophylline loadings of 0.11% and 0.18% were obtained for H-4-49 and H-9.5-54, respectively. These low loading levels are due to the low solubility of theophylline in water (0.83% (4) and the low water swelling of these networks, i.e., ~ 10%. Efforts are underway to increase the extent of loading by increasing the loading time, temperature, and using ethanol as a loading solvent.

Figure 5 shows the fraction of drug released as a function of time for networks H-4-49 and H-9.5-54. Theophylline-loaded systems provide sustained delivery. The release kinetics and mechanism were analyzed by $M_t/M_\infty = kt^n$, where M_t is the amount of drug released at time t, M_∞ is the total amount released, k is a constant that is characteristic of the system, and n an exponent characteristic of the mode of transport.(2) As shown in Figure 6, n can be obtained from the slopes of $\ln(M_t/M_\infty)$ versus ln t plots. Release from H-4-49 and H-9.5-54 gives n = 0.72 and n = 0.65, respectively, indicating anomalous diffusion for both networks. According to these results the release kinetics may be controlled by the M_n of the MA-PIB-MA and the PIB content combined. Further research is in progress in this direction.

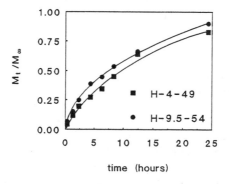

Figure 5. Release of theophylline as a function of time from HEMA-1-PIB amphiphilic networks.

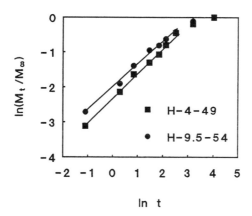

Figure 6. The $\ln(M_t/M_\infty)$ versus ln t for release of theophylline from PHEMA-1-PIB amphiphilic networks.

Acknowledgment

Support by NSF grant DMR-89-20826 is gratefully acknowledged.

Literature Cited

1. Iván, B.; Kennedy, J. P.; Mackey, P. W. previous paper in this volume.
2. Peppas, N. A.; Korsmeyer, R. W. In *Hydrogels in Medicine and Pharmacy;*
    Peppas, N. A., Ed., CRC Press: Boca Raton, Florida, Vol. III; 1987, pp. 109-
    135.
3. Greene, T. W. *Protective Groups in Organic Synthesis;* John Wiley & Sons:
    New York, NY, 1981; p. 40.
4. *The Merk Index;* Windholz, M.; Budavari, S.; Stroumtsos, L. Y.; Fertig, M. N.,
    Eds.; Ninth Edition; Merk and Co., Inc.: Rathway, N. J.,1976.
5. Okano, T.; Nishiyama, S.; Shinohara, I.; Akaike, T.; Sakurai, Y.; Kataoka, K.;
    Tsurta, T. *J. Biomed. Mat. Res.,* **1981**, *15,* 393.
6. Shimada, M.; Miyahara, M.; Tahara, H.; Shinohara, H.; Okano, T.; Kataoka,
    K.; Sakurai, Y. *Polym. J.,* **1983**, *15,* 649.
7. Shimada, M.; Sugiyama, N.; Shinohara, I.; Okano, T.; Kataoka, K.; Sakurai, Y.
    Eur. Polym. J., **1983**, *19,* 929.
8. Okano, T.; Uruno, M.; Sugiyama, N.; Shimada, M.; Shinohara, I.; Kataoka, K.;
    Sakurai, Y. *J. Biomed. Mat. Res.,* **1986**, *20,* 1035.

RECEIVED March 19, 1991

Chapter 20

Cross-Linked Polyacid Matrices for Oral Drug Delivery

Robert Z. Greenley[1,3], Hossein Zia[5], Joel Garbow[2,3], and Robert L. Rodgers[4,5]

[1]Chemical Sciences, [2]NMR Consortium, [3]Monsanto Corporate Research, Monsanto Company, St. Louis, MO 63167
[4]Department of Pharmacology and Toxicology, [5]College of Pharmacy, University of Rhode Island, Kingston, RI 02881

Concentrated solutions of many therapeutic agents can be imbibed into a crosslinked polyacid powder. After removal of the solvent, the drug-polymer matrix may be administered orally. The polyacid swells minimally under gastric conditions - retaining the drug - and then is highly swollen by the basic intestinal medium, allowing the drug to be dissolved away from the matrix. These matrices, with insulin as a model agent, have been characterized by, in vitro release curves, drug-polymer interaction studies and in vivo animal studies. ^{13}C REDOR NMR experiments were also run to define the dispersion of the agents in the polyacid resin.

The promise of the isolation and production of therapeutic polypeptides and proteins demands that for treatment of a chronic disease state an oral delivery system be developed which will protect these valuable agents from the hostile gastric environment. Subsequently, the drugs will have to be completely released in the intestine, preferably in a state that will enhance their rapid dissolution and transport across the gut wall minimizing interaction with intestinal proteases.

The design of a proper delivery system requires a knowledge of the G.I. tract (1). The nature of the gastric acidic and enzymatic medium has been elucidated. More recently Davis (2) and Harris (3) have studied the rate of emptying of the stomach. Dressman (4) has clinically followed pH variations in both the empty stomach and after a meal finding that sinusoidal pH reductions occur during mastication of solid food, whereas the duodenum maintains a relatively constant pH during introduction of the chyme.

In 1987 Saffran and Neckers (5) disclosed a coating for a capsule containing the drug that appears to satisfy a need for oral insulin delivery when tested in animals. In this case, a solution of "crosslinked" polymer is coated onto the capsule.

0097–6156/91/0469–0213$07.00/0

When the capsule reaches the ileo-cecal junction, the native bacteria cause rupture of the azo crosslink, allowing dissolution of the polymer and capsule material, releasing the insulin.

Our approach to a delivery system is also based on a crosslinked system but the polymer crosslinks are maintained, rendering the nontherapeutic portion of the formulation insoluble and thus it cannot be absorbed by any portion of the G.I. tract. A polyacid, such as poly(acrylic acid), dissolves in water. When it is neutralized with a base, it is more readily dissolved. If the polyacid is crosslinked by running the polymerization in the presence of five to ten mol percent of a diacrylate, then it is no longer water soluble and is only slightly swollen by water. When the crosslinked polyacid is neutralized by base, a polyanion is formed. In this case, the polymer swells significantly in water to relieve the repulsions of the adjacent anions. If a drug is contained within the polymer matrix powder, it will be protected (except for any that is on the surface of the polymer particles) from the gastric environment and available for dissolution by the intestinal medium after neutralization of the acid functions by duodenal bicarbonate. If other polyacids, such as poly(methacrylic acid) or their copolymers are used, the swelling characteristics of the resins will be changed according to the hydrophobicity of the polymer backbones (Figure 1).

Methods and Materials

Polymer Synthesis. A solution of 30 mL of (meth)acrylic acid, 5 mol% triethyleneglycol di(meth)acrylate and 0.3 g $K_2S_2O_8$ in 300 mL deionized water was stirred and heated at 95°C under N_2 for two hours. The gelatinous mass was vacuum dried at 50°C for 48 hours. The polymer was then ground in a Micro-Mill. The portion passing through a 40 mesh screen (<425m) was extracted three times with 80-90°C water and then redried.

Drug Matrix. To 1.0 g of resin was added a 2.5 to 5.0 mL portion of a drug concentrate in water or ethanol. The slurry was mixed for one hour, at which time most of the solution had been absorbed by the powder. The solvent was removed by lyophilization (water) or by vacuum drying at 35°C (ethanol). The resultant matrix may be used as is, reground or tableted.

HPLC. Most of the release curves were developed using High Pressure Liquid Chromatography on a VYDAC 218TP54 "Protein and Peptide C_{18}" hydrophobic interaction column. The mechanical system consisted of a Spectra- Physics SP8800 pump and SP8500 mixer. The detector was a Kratos Spectroflow 757 which was attached to a Spectra-Physics SP4290 integrator/plotter. The detector was set at the appropriate wavelength for the agent being released. All elution solvents were HPLC grade.

Experimental NRM. Solid-state ^{13}C NMR spectra were collected on a homebuilt spectrometer which operates at a proton resonance frequency of 127.00 MHz, corresponding to a carbon resonance frequency of 31.94 MHz, and employs a

Nicolet 1280 computer and 293B pulse programmer. Samples were spun at the magic angle (54.7°) with respect to the static magnetic field in a double-bearing rotor system at rates of 3 kHz. A more detailed description of the magic-angle spinner can be found elsewhere (*23*). REDOR spectra were collected following [1]H-^{13}C spin-lock contacts of 2 msec with matched, 50 kHz rf fields. High-power proton dipolar decoupling (75-110 kHz) was used in all experiments.

Rotational-echo double-resonance (REDOR)(*18,19*) is a new solid-state NMR technique which is sensitive to through-space carbon-nitrogen interactions between selectively ^{13}C and ^{15}N-enriched sites separated by up to ~5A (20-22). The parameter directly measured in a REDOR experiment is the heteronuclear dipolar coupling constant D_{CN}, which is in itself proportional to the inverse third power of the internuclear distance, r_{CN}. It is this dependence on $(r_{CN})^{-3}$ which accounts both for REDOR's ability to accurately measure short distances and its insensitivity to longer-range interactions. As a technique which can probe, in detail, intermolecular interactions over a distance range of 5A, REDOR is well suited to studying the distribution of small selectively-labeled molecules in polymer delivery systems.

In normal cross-polarization magic angle spinning (CPMAS) NMR experiments, the [13]C-^{15}N dipolar coupling is removed by magic-angle spinning and does not effect the spectrum of the observed nucleus. In REDOR, π pulses applied synchronously with the rotor serve to reintroduce this heteronuclear dipolar coupling. REDOR is performed as a difference experiment. The basic pulse sequence for the ^{13}C observe version of the experiment is shown in Figure 2. On alternate scans of the experiment, the π pulses on the nitrogen channel are either applied or omitted. Signals from alternate scans are accumulated and Fourier-transformed separately. When applied, the [15]N π pulses cause a partial dephasing of ^{13}C signal due to the [13]C-^{15}N dipolar coupling, leading to a loss in ^{13}C signal intensity. The magnitude of this signal loss, ΔS, is determined by subtracting spectra from the experiments with and without [15]N π pulses and depends on the strength of the heteronuclear dipolar coupling. When normalized by the magnitude of the 'parent' signal from the labeled ^{13}C position, S_0, this REDOR difference signal serves as a measure of this heteronuclear coupling and, thereby, carbon-nitrogen internuclear distance.

The magnitude of the normalized REDOR difference signal, $\Delta S/S_0$ depends upon the strength of the heteronuclear dipolar coupling and on two parameters which are controlled experimentally: (1) spinning speed (v_R) and (2) number of rotor cycles of [13]C-^{15}N dipolar coupling evolution (N_C). The expressions which relate these quantities have been derived (*19*) and can be evaluated with the aid of a computer. For a sample of small. isotopically-enriched molecules distributed within a polymer matrix, however, there is, in all likelihood, a distribution of distances between labeled carbon and nitrogen sites. Depending upon the detailed distribution of the small molecules, a labeled carbon site may in fact be coupled to more than one ^{15}N spin. A detailed, quantitative analysis of REDOR data from such a sample requires knowledge of the relevant carbon-nitrogen distances, and, for carbons interacting with more than one nitrogen, the relative orientation of the different C-N

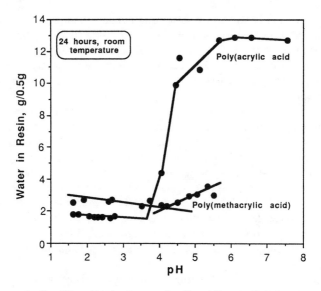

Figure 1. Swelling titration curves for 5 mol% crosslinked polyacids.

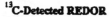

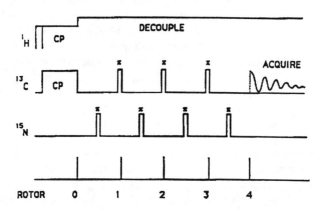

Figure 2. Pulse sequence for ^{13}C-serve REDOR NMR. This sequence differs from the original REDOR pulse sequence (ref. 18) in that π pulses alternate between ^{13}C and ^{15}N r.f channels. On alternate scans of the REDOR experiment, the ^{15}N π pulses are either applied or omitted. This figure illustrates that the REDOR pulse sequence with four rotor periods of ^{13}C-^{15}N dipolar-coupling evolution ($N_C = 4$) NC can be increased (in increments of two) by adding rotor periods and pairs of ^{13}C and ^{15}N π pulses between the end of the cross-polarization preparation and the start of data acquisition.

internuclear vectors. No such analysis will be attempted here. Rather, we will take the quantity $\Delta S/S_o$ as a measure of the *average* ^{13}C-^{15}N dipolar coupling, with an increase in its magnitude reflecting an increase in the degree of intermolecular interaction between labeled small molecules. As we will show, such a semi-quantitative treatment of the REDOR data permits much to be learned about the distribution of these small molecules within the polymer matrix.

Diabetic Rats-Phase I. Laboratory rats (CD strain, 250-300g, male) were made diabetic by a single injection of streptozotocin (STZ), 50 mg/kg, into the tail vein. Nondiabetic controls received an equal volume of citrate buffer. Twenty-four hours after the STZ injection, each rat was individually housed for urine collection. The appearance of glucose in the urine (Ames test strips) and a predictable weight loss or depression of the growth curve were taken as confirming evidence of diabetes.

The diabetic rats were treated with 18 IU of bovine insulin imbibed into polyacid resins b.i.d. orally using 1 cc syringes and gavage tubes. After 14 days of treatment the rats were sacrificed about 1.5 hours after the last dose. Blood samples were taken and assayed for immunoactive insulin activity (Amersham-Searle RIA kit) and serum glucose levels (glucose oxidase colorimetric assay, Sigma 510 Glucose Kit).

Diabetic Rats-Phase II. This protocol was similar to that in phase I except that 65 mg/kg STZ was employed, 54 IU bovine insulin in the polyacid resin was utilized in each dose and the treatment period was for nine days. Groups of rats (four per group) were then sacrificed at 30, 90, 165 and 255 minutes after the last dosing.

Normal Rabbits. Six male, white rabbits (2.5 - 3.0 kg) were housed individually. Animals were fasted overnight for 16 hours (with access to water) prior to each experiment to reduce the gastrointestinal content and absorption variability. After treatment with either a control dose or experimental insulin in poly(acrylic acid) resin dose, a one week washout period was required before the next experiment. The protocol called for blood samples to be taken from an indwelling ear catheter at -1, -.5, -.25, +.5, +1, +1.5, +2, +3, +4, +5 and +6 hours. Serum glucose levels were determined by an oxidase colorimetric method using the Sigma 510 Glucose Kit.

Diabetic Rabbits. Nine rabbits were rendered diabetic by an intravenous (iv) injection of 45 mg/kg alloxan. Urine and blood samples were tested one day before and two days after alloxan treatment for estimation of glucose content by chemstrip UGK and bG, respectively. The experimental protocol was the same as that for normal rabbits.

Angiotensin II. Male Sprague-Dawley rats (300-400 g) were administered either free angiotensin II (A-II) (Sigma product #A9525, human synthetic A-II, Asp-Arg-Val-Tyr-Ile-His-Pro-Phe) in 0.9% NaCl solution by subcutaneous injection, or A-II impregnated resin by gastric intubation. The resin was suspended in distilled water such that the rat received no more than a 0.5 mL total suspension volume. The controls in each group were administered equal volumes of saline or water, respectively. At various times after A-II administration, arterial blood pressure was determined by the tail-cuff method (26), at an ambient temperature of 32°C. In a subgroup of animals, serum A-II activity was measured, from blood samples obtained during the maximum pressure response to A-II injection (3-5 minutes), by radioimmunoassay (Peninsula Laboratories, Belmont, CA).

Results

In vitro Studies - Release Rates. To test the concept, an initial in vitro screening of four agents imbibed into four polyacid resins of varying degrees of crosslinking was performed, as summarized in Table I. The first three agents were of decreasing molecular weight and the last, 5-aminosalicylic acid was of interest since it is apparently topically active in the treatment of ulcerative colitis but cannot survive a gastric environment. Slurries were stirred first in a synthetic gastric medium (6) for one hour, the solids centrifuged and the supernatant assayed for released agent. The solids were then redispersed in a synthetic intestinal fluid (7) to which sufficient isotonic bicarbonate solution was added to neutralize the resin acid functions. After stirring for two hours the supernates were again assayed by ultraviolet spectroscopy for released agent. From these "one-shot" gross release results the five percent crosslinked acrylic and methacrylic acid resins were selected for further study.

HPLC reverse phase procedures were established to follow the continuous release rates of a variety of agents from the two resins. Also a USP standard release test procedure (8) was used. Because of its ease of detection at the higher ultraviolet wave lengths, bovine insulin was used as the model delivery agent.

Because of the fortuitous timing for sampling the release in the intestinal medium, the "one-shot" approach suggested that the methacrylic acid resin would be better as a delivery matrix for insulin but, the acrylic acid system was shown to be a better choice (Figures 3a,b). Even though there was a rapid, initial release in gastric fluid using the latter system, possibly due to insulin adsorbed on the surfaces of the particles, no further release occurred. In the methacrylic acid case the release was continuous although slower in the gastric medium than in the intestinal fluid.

The method of polymerization also affects the way insulin is released from the matrix. Conventionally (9), the polymerization of acrylic acid with a crosslinking agent is carried out in a saturated aqueous solution of $MgSO_4 \cdot 7H_2O$. This technique allows one to remove the solid polymer easily from the polymerization medium whereas, when the polymerization is carried out in the absence of a coagulant - the procedure used in this work - a difficultly handled, highly swollen gel results. Interestingly, the gastric release rates for resins prepared in these two systems are quite different (Figure 4). Crosslinked poly(acrylic acid) prepared at 30°C vs. the normally employed temperature of 95°C using a redox initiating system gave gastric insulin release similar to that of the polymer prepared in the saturated salt medium.

When the five mol percent crosslinked poly(acrylic acid) release curve was compared (Figure 5) to that for a very lightly crosslinked (less than one percent)

Table I. Percent Extraction of Imbibed Agents from One Gram of Polyacid Resins Under Synthetic Gastric (50 mL/1hour) Followed by Intestinal (120 mL/2 hours) Conditions

PROTECTING RESIN	Human serum albumin (50mg/g)		Bovine Insulin (50mg/g)		Tyrosine-Alanine Dipeptide (40mg/g)		5-Amino-Salicylic Acid (40mg/g)*	
	gast	int	gast	int	gast	int	gast	int
Acrylic Acid								
2% crosslinked	0	64	0	55	79	10	77	24
5% crosslinked			10	78			87	14
10% crosslinked			5	73				
Methacrylic Acid								
3% crosslinked	0	75	4	28	10	41	12	64
5% crosslinked			1	43			69	31
10% crosslinked			2	31				
Ethylene/Maleic Anhydride Copolymer								
5% crosslinked	0	96	---	100	92	20	78	32
Methyl Vinyl Ether Maleic Anhydride Copolymer								
5% crosslinked	0	17	2	10	31	36	49	41

* milligrams of agent imbibed into one gram of resin

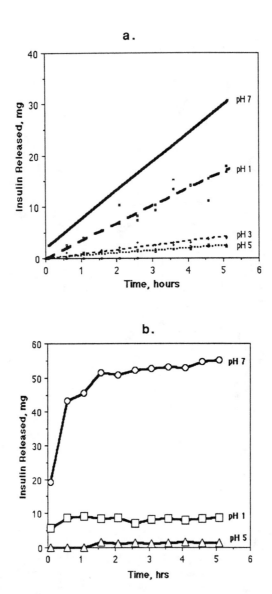

Figure 3. HPLC release curves for insulin imbibed into (a) poly(methacrylic acid) and (b) poly(acrylic acid).

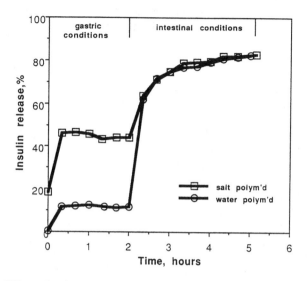

Figure 4. Effect of polymerization conditions on the release characteristics of the insulin/poly(acrylic acid) resin system.

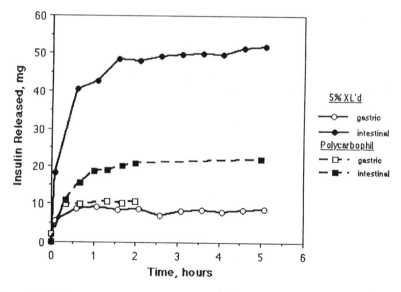

Figure 5. Release of insulin from lightly crosslinked (Polycarbophil) and 5 mol% crosslinked poly(acrylic acid).

polyacid, Polycarbophil, the rates of release of insulin in gastric fluid were similar. However, only one third of the insulin was available from the Polycarbophil under basic, intestinal conditions. This difference was attributed to the very highly swollen gel formed by Polycarbophil after neutralization with bicarbonate. Diffusion of solution out of the gel into the supernate would be very slow from the interstices of this sticky mass.

One-to-one random copolymers of acrylic acid with either hydroxyethyl acrylate (a hydrogel model) or methyl acrylate failed to protect insulin from release under gastric conditions (Figure 6). In the case of the hydrogel, the expected swelling due to exposure to water occurred, releasing insulin. The behavior of the ester copolymer led to the prediction that there should be no more than about four carbon atoms per carboxylic acid group in a repeat unit of the polymers. We have not been able to disprove this hypothesis thus far.

Affinities. It became apparent that not all agents could be properly protected and then released from a given polyacid resin. In an attempt to better predict the more successful pairings, a series of experiments was run in which aqueous resin slurries were stirred with solutions of agents. The supernates were then assayed for the amount of the solubles which were not absorbed by the wet polymer powders. The results are summarized in Table II. In almost all cases the agents were readily absorbed by the acrylic acid resin even though all but insulin were found to be readily released from the polymer under gastric conditions. This affinity may be due to the fact that there is a somewhat acidic hydrogen at the two-position of the repeat unit. The rest of the matches correspond better with the increasing concentration and hydrophobicity of the substituent at the second carbon. The highlighted numbers in each section have been confirmed by release curves. No satisfactory carbon-carbon backbone polymer was found to be appropriate for any of the agents that had high calculated surface energies, such as aspirin. The minimal effect of charge as shown by the two salicylic acid experiments has been confirmed by release curves for a number of other neutralized compounds and their corresponding free acid or base state. The sodium salt of indomethacin was also included in the nonsteroidal antiinflammatory drug (NSAID) series. Apparently the resins were more acidic than the agent as the precipitated free indomethacin (except for the resin free control) was recovered at the end of the equilibration period.

Dispersion. If the technique to imbibe the agent into the polymer is examined, one might assume that a solid solution or molecular dispersion might result at agent/drug ratios below certain levels. These levels would probably depend upon the compatibility of the two materials, the molecular weight of the agent and perhaps the effect of the agent containing solvent (water or ethanol) on the resin. To investigate this possibility, a relatively new NMR analysis has been employed.

Solid-state NMR can be applied equally well to both crystalline and amorphous solids (10-17), and can be used to distinguish and quantify the crystalline and amorphous components of multi-component solids. Here we describe the

Table IIa. Percent Absorption of Drugs by Polyacid Resins [50 mg agent/gram resin]

Drug	Acrylic Acid	1/1-Acrylic Acid/ Methacrylic Acid Copolymer	Methacrylic Acid	1/1-Ethacrylic Acid/ Methacrylic Acid Copolymer	Ethacrylic Acid
Bovine Insulin	<u>83</u>	17	35	0	0
Verapamil Hydrochloride*	74	36	33	<u>53</u>	4
Erthromycin	71	34	35	<u>64</u>	2

* relative estimates due to partial degradation by the acidic resins

Table IIb. Percent Absorption of NAIDs by Polyacid Resins [50 mg agent/gram resin]

Drug	Acrylic Acid	1/1-Acrylic Acid/ Methacrylic Acid Copolymer	Methacrylic Acid	1/1-Ethacrylic Acid/ Methacrylic Acid Copolymer	Ethacrylic Acid
Na+ Ibuprofen	79	81	72	44	15
Na+ Naproxen	94	94	85	62	89
Na+ Salicyclic Acid	<u>4</u>	8	9	2	7
Salicyclic Acid	<u>5</u>	10	11	10	12

application of a new solid-state NMR technique, REDOR, to study the distribution of small molecules within an amorphous polymer matrix.

Figure 7 shows REDOR ^{13}C NMR spectra of a sample of poly(acrylic acid) (PA) imbibed with equal amounts of [3- ^{13}C]Ala and [^{15}N]Ala (50:1:1, by weight). These spectra were collected using the pulse sequence of Figure 6 with 30 rotor periods of ^{13}C-^{15}N dipolar coupling evolution (N_C=30). Figure 7 (bottom) is the full echo spectrum (S_0) of the PA/alanine sample and shows ^{13}C resonances due to both the labeled alanine methyl carbon (20 ppm) and natural-abundance carbons in alanine and the PA polymer. The relatively broad lines observed in this spectrum are typical of those seen in the solid-state ^{13}C NMR spectra of polymers (3,4). In solid polymers, variations in both intramolecular chain conformation and intermolecular chain packing create heterogeneous local environments which produce distributions of isotropic chemical shifts. It is these distributions of isotropic shifts which dominate the line widths in many polymer systems. As demonstrated by the line width of the alanine methyl-carbon signal, the polymer transfers this heterogeneity of local environment to the imbibed small molecule, producing a comparable distribution of chemical shifts for it. Figure 7 (top) is the REDOR difference spectrum (ΔS) for this sample, produced by subtracting spectra collected with and without ^{15}N π pulses. The $\Delta S/S_0$ value of 0.08 for the methyl carbon reflects a relatively weak average ^{13}C-^{15}N dipolar coupling. Of this value, 0.014 is calculated to arise from intramolecular interactions between natural-abundance carbon (nitrogen) and ^{15}N (^{13}C) label. The remainder of this interaction is intermolecular in nature. If it is assumed each ^{13}C is coupled to only one ^{15}N, the observed $\Delta S/S0$ translates into an average carbon-nitrogen internuclear distance of 4.8A.

Figure 8 shows REDOR ^{13}C NMR spectra for a sample of PA imbibed with higher concentrations of [3-^{13}C]Ala and [^{15}N]Ala (10:1:1,by weight). Again the bottom spectrum is the echo spectrum of the entire sample and the top is the REDOR difference spectrum. As Figure 8 (bottom) shows, the alanine methyl carbon signal is split into two lines: a relatively broad line at 20 ppm and a much sharper signal at 17 ppm. The former corresponds in chemical shift with the alanine methyl-carbon signal in Figure 7: the latter with the methyl-carbon chemical shift of crystalline alanine (spectrum not shown). This spectrum indicates that some of the alanine has apparently phase separated into small homogeneous domains, giving rise to signal whose shift and line width are like that of alanine itself. This interpretation is confirmed by the REDOR difference spectrum of Figure 8 (top) which shows that $\Delta S/S_0$ for the sharp-line methyl-carbon signal is much larger than that of the broad line component, as would be expected for well-ordered crystalline-like domains. We have studied a series of PA polymers imbibed with increasing concentrations of (equimolar) mixtures of [^{13}C]Ala and [^{15}N]Ala. TableIIIshows values of $\Delta S/S_0$ measured for each of these samples. As noted above, the spectrum of the sample containing the highest concentration of alanine displays two resolved methyl-carbon signals, whose individual $\Delta S/S_0$ values are reported in the table. For the broad- line methyl signal, $\Delta S/S_0$ increases with increasing alanine concentration.

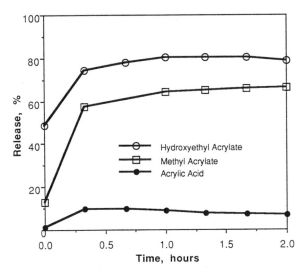

Figure 6. Release of insulin under gastric conditions from crosslinked 1/1 random acrylic acid copolymers with either hydroxyethyl acrylate or methyl acrylate.

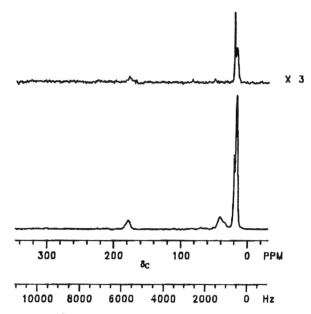

Figure 7. REDOR ^{13}C NMR spectra of poly(acrylic acid) (PA) imbibed with [3-^{13}C]Ala/[^{15}N]Ala (50:1:1 by weight). The bottom curve represents the echo spectrum of full sample (S_0); The top curve is the REDOR difference. (ΔS). Spectra were collected using the pulse sequence of Figure 2 with $V_R = 3$ kHz; $N_C = 30$.

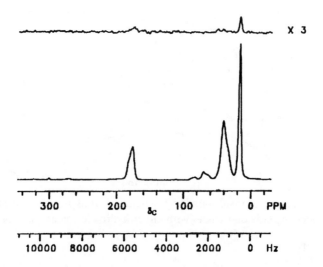

Figure 8. REDOR ^{13}C NMR spectrum of poly(acrylic acid) (PA) imbibed with
[3-^{13}C]Ala/[^{15}N]Ala (10:1:1 by weight). The bottom curve represents the echo
spectrum of full sample (S_0); the top curve is the REDOR difference. (ΔS).
Spectra were collected using the pulse sequence of Figure 2 with V_R = 3 kHz;
N_C = 30.

Table III. Poly(acrylic acid)/[3-^{13}C] Alanine Formulations

[3-^{13}C]Ala	[^{15}N]Ala	$\Delta S/S_o$
5 mg	5 mg	0.028
10 mg	10 mg	0.040
25 mg	25 mg	0.062
50 mg	50 mg	0.30 (sharp)
		0.085 (broad)

We have also examined three other pairs of amino acids: (1) [1-^{13}C]Leu/[^{15}N]Leu imbibed into poly(methacrylic acid), PMA; (2) [1-^{13}C]Gly/[^{15}N]Gly imbibed into PA and (3) [1-^{13}C]Ala/[^{15}N]Ala imbibed into PA and PMA. Because the polymers contain carboxyl carbons, carbon signals from the labeled amino acids are not chemical-shift resolved in these small molecule/polymer systems. Consequently, the natural abundance polymer carbonyl-carbon signals must be properly accounted for in interpreting REDOR results from these systems. Natural-abundance polymer carboxyl-carbon signals contribute to both S_0 (all polymer molecules) and ΔS (polymer molecules which are within 5A of [15N]-labeled alanines). Each of these contributions can be quantitatively measured. Polymer contributions to S_0 can be determined and corrected for by collecting the spectrum of natural-abundance polymer alone and then subtracting it from the spectra of each of the imbibed samples. REDOR difference signals arising from the polymer carboxyl carbons can be measured by examining samples containing [^{15}N]-labeled alanine only. Figure 9 summarizes REDOR data for the PA/[1-^{13}C]Ala/[^{15}N]Ala and PA/[1-^{13}C]Gly/[^{15}N]Gly systems. In determining each of these $\Delta S/S_0$ values, natural abundance contributions to S_0 have been subtracted, as described above. Natural-abundance contributions to ΔS have been measured for a number of samples and are small; the numbers in this table have not been adjusted for this effect. As was seen for the [3-^{13}C]Ala/[^{15}N]Ala samples, we observe a moderate increase in $\Delta S/S_0$ with increasing small-molecule concentration. For the alanine sample at highest concentration, a distinct separation of phases similar to that displayed in Figure 7 is observed. For the glycine sample, no obvious narrowing of the line is observed, even at the highest concentration.

The REDOR ^{13}C echo spectra of two different PMA/[1-^{13}C]Leu/[^{15}N]Leu samples are shown in Figure 10. Signals from the enriched carboxyl carbon of leucine are sharp at all concentrations. Apparently, even at the lowest concentration, the leucine molecules are not well dispersed throughout the polymer, but instead are clustered in small, ordered, crystalline-like domains. The doubling of the carboxyl signal is also seen for pure, crystalline leucine and is attributable to the detailed packing of the leucine molecules in the solid state.

As described above, for a heterogeneous sample containing a distribution of distances between labeled carbon and nitrogen sites, REDOR ^{13}C NMR measures an average ^{13}C-^{15}N dipolar interaction. Such is certainly the case for the amino-acid imbibed/PA model drug-delivery systems examined here. From this average interaction, however, we can learn much about the distribution of small molecules within these polymers and can begin to model these distributions. For example, the REDOR results demonstrate that intermolecular interactions between amino acid molecules are present, even at concentration as low as 2% (by weight) amino acid in PA. These interactions necessarily arise from 'clustering' of amino-acid molecules. One possible distributional model is to assume that the average size of a cluster of small molecules within the polymer is fixed (determined, perhaps, by the average size of a void between polymer chains). According to this model, as the concentration of small molecules increases, the number, but not the average size, of such clusters would increase. This leads to the prediction that $\Delta S/S_0$ in the REDOR

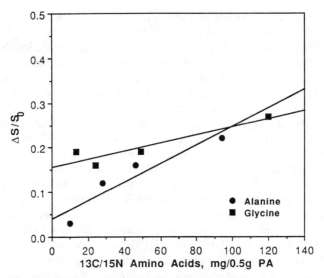

Figure 9. Differentiation between dispersions of alanine or glycine in poly(acrylic acid) resin (PA) as a function of concentration.

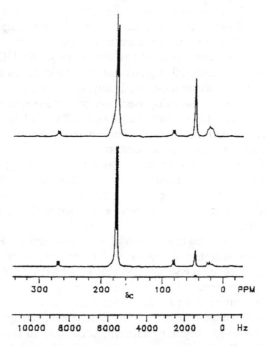

Figure 10. REDOR ^{13}C NMR echo spectra of PA imbibed with equimolar amounts of[1-^{13}C]Leu/[^{15}N]Leu: (bottom) 10:1:1, by weight; (top) 50:1:1, by weight. Spectra were collected using the pulse sequence of Figure 2 with $V_R = 3$ kHz; $N_C = 30$.

experiment should be independent of concentration. This prediction is clearly inconsistent with the REDOR data, however, which show an increasing intermolecular interaction with increasing small-molecule concentration. Instead, REDOR demonstrates that the average size of the cluster, and hence the number of neighbors with which a given molecule can interact, increases with concentration. Eventually these clusters become large and well-ordered enough that phase-separation occurs. The resultant domains of small-molecules produce sharp-line signals and have short intermolecular distances (large $\Delta S/S_0$ values) approaching those of the respective crystalline, labeled amino acids. Experiments to further refine and define this distributional model, including REDOR study of amino acids distributed in alternate polymer matrices, are in progress.

In vivo Studies with Insulin.

As it was a model compound for characterization of the in vitro characteristics of the polyacid resin delivery systems, it was also employed as the model agent for the in vivo evaluation of these systems. In the first experiment, a group of diabetic rats (50 mg/kg streptozotocin, STZ) were treated for 14 days with 18 IU insulin, b.i.d., which had been imbibed into poly(acrylic acid) (50 mg/g resin) and poly(methacrylic acid) (50 and 100 mg /g resin). The controls were a no-treatment group and a nondiabetic, untreated group. Shortly after the last dose, the animals were sacrificed and the sera assayed for glucose and insulin levels. The results, Figure 11 where each data point represents the results for a single animal, showed that there appeared to be a significant drop in serum glucose for 50 mg/g resin formulations in each of the acids. The lack of effect for the 100 mg/g poly(methacrylic acid) resin might be due to any of a number of reasons: (1) at this loading the postulated monomolecular dispersion is not achieved thus reducing the rate at which the "crystalline" dissolves; (2) the presence of contiguous insulin molecules which are dissolved by the gastric fluids, forming channels through which the fluids can reach and dissolve or enzymatically degrade the more deeply embedded protein molecules; (3) polymerization of the insulin monomer may occur more readily at these higher concentrations.

In an effort to better characterize these encouraging results, an experiment was run to determine the average time after dosing that the rise in insulin levels and drop in serum glucose occurred. To emphasize the effect, even higher levels of STZ (65 mg/kg) and insulin (54 IU, b.i.d.), using the 50 mg/g poly(acrylic acid) formulation were evaluated. After nine days of treatment, groups of rats were sacrificed over four time intervals after the last dose. Figures 12a and b show that at about two and a half hours after dosing a major lowering of glucose occurred ($p = 0.01$) with a concurrent, apparent increase in serum insulin ($p = 0.19$).

To improve the reliability of the in vivo results, new studies were initiated in larger animals with a sufficient blood supply to act as their own controls. In these experiments normal rabbits were fitted with an indwelling ear catheter which facilitated sampling of the animal's blood at frequent intervals. Two controls were used in these studies: a positive control group in which a 0.25 IU/kg dose of

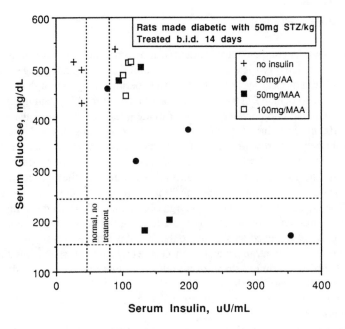

Figure 11. Serum assays of diabetic rats after 14 days of treatment with 18 IU of bovine insulin, b.i.d., in polyacrylic and polyacrylic acid resins.

a.

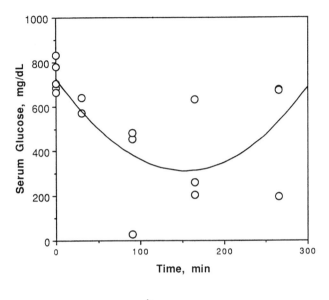

b.

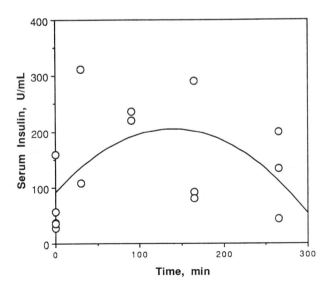

Figure 12. Minimum serum glucose levels (a) and maximum serum insulin levels (b) as a function of time after the last treatment with 54 IU insulin, b.i.d., for nine days.

crystalline bovine insulin was administered subcutaneously and a neutral control group which was treated with an oral dose (gavage) of 25 IU/kg of crystalline bovine insulin which had been dry blended with an appropriate amount of poly(acrylic acid) resin. The poly(acrylic acid) resin formulation which contained 50 mg insulin/g resin was orally administered at two levels: 10 and 25 IU/kg. As shown in Figure 13, the data are normalized to their average baseline (pretreatment) serum glucose levels and shown as a percent glucose reduction. Six animals were in each of the control groups, five in the 10 IU oral dose and four in the 25 IU oral dose groups. The dry blend had no effect on the serum glucose level while the subcutaneous injection had a dramatic and rapid effect. The major lowering of the glucose levels again occurred about two and one half hours after dosing. The area under the curve from zero to six hours for the 25 IU p.o. dose is approximately half that of the 0.25 IU subcutaneous dose. This implies a 0.5% bioavailability of solid insulin when administered orally using this resin imbibed approach.

When diabetic rabbits (24) were treated with 50 IU of bovine insulin imbibed at 50 mg/g poly(acrylic acid) (Figure 14) no reduction in serum glucose over that achieved by the dry blend control could be detected. Pretreatment of the animals with oral doses of either a penetration enhancer, sodium taurocholate, or a protease inhibitor, aproteinin, failed to improve the insulin activity. One possible explanation for this unexpected lack of activity might be that the diseased animals exhibit impaired ileal absorption of fluids (25).

In Vivo Studies with Angiotensin II. Preliminary studies were carried out in order to provide indirect evidence for gastrointestinal absorption of angiotensin II (A-II) after oral administration of A-II impregnated resin particles. Subcutaneous injections of A-II caused a dose-dependent increase in systolic arterial pressure, over a dose range of 25-400 mg/kg, confirming previous observations (27,28). The maximal pressure response to 200 mg/kg A-II was associated with significantly elevated serum A-II activity (Figure 15). The elevated basal A-II activity of the control group may have been attributable to antiserum specificity, cross-reacting A-II fragments, or other plasma components (29,30). In this study, no attempt was made to remove interfering substances prior to radioimmunoassay. In contrast, intragastric administration of A-II impregnated resin particles, at a dose of 5 mg/kg A-II (50 mg A-II per g resin), had no apparent effect on arterial pressure up to three hours post administration (Figure 16).

The results obtained from A-II injected animals (Figure 15) confirmed that the peak arterial pressure response is a reliable indirect indicator of A-II absorption (27,28). On this basis it is very unlikely that oral administration of A-II-impregnated resin (Figure 16) resulted in any significant absorption, even at an A-II dose which was 25X higher than the maximally effective subcutaneous dose. As in the insulin studies, the detectable response was observed about two and one-half hours after dosing.

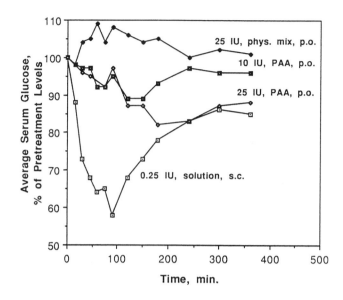

Figure 13. Normalized reduction of serum glucose in normal rabbits by oral administration of bovine insulin imbibed into a poly(acrylic acid) matrix.

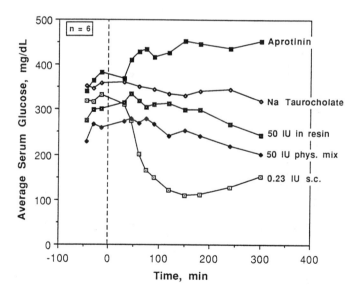

Figure 14. Effect of oral delivery of an insulin/polyacid matrix to diabetic rabbits on serum glucose, with and without adjuvants (sodium taurocholate, aprotinin).

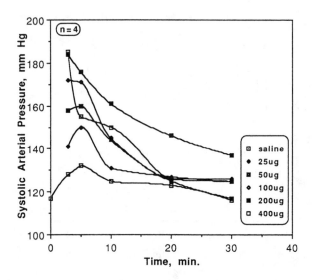

Figure 15. Dose response of rats to Angiotensin II administered iv.

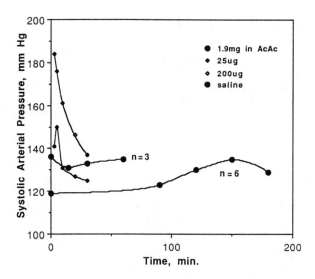

Figure 16. Oral administration of Angiotensin II imbibed into poly(acrylic acid) resin to rats.

Discussion

These delivery systems, which are based on completely absorbing (imbibing) a concentrated solution of drug into a crosslinked polyacid powder and then removing the solvent, initially seemed to be quite simple and universal in concept. As usual, the more the information that was developed, the more information was needed to decipher the anomalies. It is now apparent that these polyacid resins with carbon-carbon backbones are not at all suitable for the gastric protection and subsequent intestinal release of drugs with high solubility parameters, such as acetyl salicylic acid. Also, the effect of the charge on the drug, as seen in the case of aspirin, is minimal. We have also tested some amine containing drugs, which are not discussed in this article, in these polyacid matrices and found that there is no significant difference in the release characteristics when the drug was imbibed as either the free base or as its hydrochloride salt.

The better drug/polyacid resin systems as developed offer excellent protection for the drug from the gastric environment or for the stomach lining from the drug. Unlike the thin, soluble enteric coatings used to coat tablets and granules, this crosslinked polymer matrix is insoluble and cannot be absorbed by any portion of the G.I. tract. There is some evidence that molecular dispersions or solid solutions result from some drug/polyacid formulations. The existence of such a dispersion should improve the rate of dissolution and hence bioavailability of the drug. Although no research has been done to prove the point, these solid solutions should provide outstanding stabilization for racemizable centers, other tertiary structures and minimize bacterial degradation of substrates.

Acknowledgments

We wish to thank T. M. Brown for his excellent development of the myriad of HPLC procedures required to characterize these and many other drug/polymer systems; C. E. Vogt for picking up the HPLC development program and for his assistance in monomer synthesis; M. Christie for his excellent implementation of the angiotensin II study; L. A. Luzzi for his guidance and P. S. Glaspie and D. Forster for their willingness to back this program both spiritually and financially.

Literature Cited

1.   Johnson, L.R. *Gastrointestinal Physiology*; The C. V. Mosby Co., St. Louis, MO, 1985, third edition,.
2.   Davis, S.S. *STP Pharma* **1986**, *2,* 1015.
3.   Fell, J. T., Harris, D., Sharma, H. L., and Taylor, D. C. *Polym. Preprints,* Div. of Polym. Chem. Inc., ACS, **1987**, *28,* 145.
4.   Dressman, J. B., Berardi, R. R., Dermentzoglou, L. C., Russell, T. L., Schmaltz, S. P., Barnett, J. L. and Jarvenpaa, K. M. *Pharmaceutical Research,* **1990**, 7, 756.

5.   Saffran, M., Kumar, G. S., Savariar, C., Burnham, J. C., Williams, F., and Neckers, D. C. *Science,* **1986**, *233,* 1081.

6.   *U.S. Pharmacopeia,* **XXI**, Rockville, MD, 1985, p. 1424.

7.   op. cit., p. 1424.

8.   op. cit., p. 1244.

9.   Ch'ng, H. S., Park, H., Kelly, P., and Robinson, J. R. J. *Pharmaceutical Sciences,* **1985,** *74,* 399.

10. Fyfe, C. A. *Solid State NMR for Chemists,* CFC Press, Guelph (1983).

11. Mehring, M. *High Resolution NMR in Solids,* Springer-Verlag; NY (1983).

12. *High Resolution NMR Spectroscopy of Synthetic Polymers,* Komoroski, R., Ed.; VCH:Deerfield Beach 1986.

13. Fedotov, V. D. and Schneider, H. *Structure and Dynamics of Bulk Polymers by NMR-Methods,* Springer- Verlag: Heidelberg 1989.

14. Pines, A., Gibby, M.G. and Waugh, J. S. *J. Chem. Phys.,* **1973,** *59,* 569.

15. Andrew, E. R., Bradbury, A. and Eades, R. G. *Nature,* **1958,** *182,* 1659.

16. Lowe, I. *Phys. Rev. Lett.,* **1959,** *2,* 285.

17. Schaefer, J. and Stejskal, E. O. *J. Am. Chem Soc.,* **1976,** *98,* 1031.

18. Guillion, T. and Schaefer, J. *J. Magn. Reson.,* **1989,** *81,* 196.

19. Guillion, T. and Schaefer, J. *Adv. Magn. Reson.,* **1989,** *13,* 55.

20. Marshall, G. R., Buesen, D. D., Kociolek, K., Redlinski, A. S., Leplawy, M. T., Pan, Y., and Schaefer, J. *J. Am. Chem. Soc.,* **1990,** *112,* 963.

21. Garbow, J. R. and McWherter, C. A. *10th International Biophysics Congress,* Whistler, B. C., Canada, July, 1990.

22. Pan, Y., Guillion, T., and Schaefer, J. *J. Magn. Reson.,* in press.

23. Schaefer, J., Garbow, J. R., Stejskal, E. O., and Lefalar, *J. Macromolecules,* **1987,** *20,* 127.

24. Bell, R.H. and Hye, R.J. *J. Surgical Research,* **1983,** *35,* 433.

25. Chang, E. B., Bergenstal, R. M., and Field, M. *J. Clin. Invest.,* **1985,** *75,* 1666.

26. Bunag, R. D. *J. Appl. Physiol.,* **1973,** *34,* 279.

27. Yamaguchi, K., Mamoru, K., and Hama, H. *Am. J. Physiol.* **1985,** *248,* R249.

28. Osborn, Jr., J. W., Skelton, M. M., and Cowley, Jr., A. W. *Am. J. Physiol.,* **1987,** *252,* H628.

29. DeSilva, P. E., Husain, A., Smeby, R. R., and Khairallah, P. A. *Anal. Biochem.* **1988,** *174,* 80.

30. Herman, K., Ganten, D., Unger, T., Bayer, C., and Lang, R. E. *Clin. Chem,* **1988,** *34,* 1046.

RECEIVED March 19, 1991

Chapter 21

Gastric Retention of Enzyme-Digestible Hydrogels in the Canine Stomach under Fasted and Fed Conditions

Preliminary Analysis Using New Analytical Techniques

Waleed S. W. Shalaby[1], William E. Blevins[2], and Kinam Park[1,3]

[1]School of Pharmacy and [2]School of Veterinary Medicine, Purdue University, West Lafayette, IN 47907

A new set of imaging techniques were used to observe the gastrointestinal transit of enzyme-digestible hydrogels in dogs. Hydrogels loaded with sodium diatrizoate/diatrizoate meglumine were detected using radiographic and fluoroscopic imaging. Swelling, propulsion, and retropulsion of hydrogels in the stomach, as well as gastric tissue-gel interactions were monitored in real-time through the use of ultrasound imaging. Combined use of ultrasound imaging and radiographic or fluoroscopic imaging allowed us to readily locate the gel and monitor its movement. Gastric retention for 24 h under fasted conditions was achieved with hydrogels which underwent minimum deformation in response to the peristaltic wave activity. Preliminary data indicate that long term hydrogel retention in the fasted stomach can be achieved if the gel size and integrity is controlled.

Recent advances in conrolled drug delivery have made it possible to deliver drugs at any desired rate for a prolonged period of time ranging from days to months (*1,2*). Despite such advances, the development of oral controlled dosage forms has been slow primarily due to the variability in the gastrointestinal (GI) transit time. The transit time of an oral dosage form from mouth to caecum varies from 3 h to 16 h depending on the state of the stomach (*3-6*). Consequently, drug concentrations in the blood may only be maintained for short and variable periods of time regardless of the controlled release properties of the device. Because of this problem, true once-a-day oral drug delivery has yet to be achieved. The primary goal in the design of oral controlled release dosage forms is to control the GI residence time of the device by overcoming the physiological barrier or barriers that contribute to the wide variation in the GI transit time. In recent years, it has generally been agreed that the gastric

---

[3]Corresponding author.

0097–6156/91/0469–0237$06.00/0

emptying time largely controls the residence time of a dosage form in the upper GI tract (7). For this reason, a great deal of interest has been focussed on controlling the gastric emptying time of the dosage form to prolong its GI residence time and improve therapeutic efficacy. Previous attempts to alter the size, shape, density, and surface properties of oral controlled dosage forms, however, had limited success in controlling the GI residence time (8-13).

Gastric motility is associated with either fed state activity or fasted state activity (14). To achieve long-term oral drug delivery, it is desirable to control the gastric residence time of the dosage form under both fed state and fasted state conditions. In the fed state, the gastric emptying of drug delivery systems will vary depending on the size of the device (3,10,12), the type of meal (3,5), and the frequency of feeding (6). In general, when the size of the device is larger than 5 mm in diameter, it will be retained in the stomach until the digestive contents of the stomach are emptied (12,15-17). The gastric residence time of the device is likely to be shortest in the presence of an acaloric liquid instillate and longest in the presence of a highly caloric solid meal (3,5,18-20). Although there is still debate regarding the mechanism of solid-liquid partitioning in the stomach (21,22), gastric retention of a dosage form for 12-16 h may be achieved with the frequent administration of food (6); however, this approach may be impractical from a patient compliance standpoint.

In the fasted state, GI motility is characterized by a cyclically recurring, 4-phase myoelectric complex known as the interdigestive myoelectric motor complex (IMMC) (23,24). The gastric retention time of many nondigestible materials is largely controlled by phase III of the IMMC (20). In the canine stomach, phase III activity occurs when 100% of the pacesetter potentials originating in the greater curvature of the orad corpus (25) are accompanied by action potentials corresponding to peristaltic contractions (26). Under maintained fasted conditions, phase III activity occurs approximately every 100 min. Phase III activity can be immediately abolished by the administration of liquids or solids and will remain absent depending on the quantity and caloric value of the food. The duration of absence, however, is more likely dependent on the physicochemical properties of the ingesta (23,27). When phase III activity resumes, it invariably begins distal to the stomach. Because of its recurring property and high propulsive efficiency, phase III of the IMMC is the most crucial physiological barrier in the GI tract that must be overcome in order to control and prolong the GI residence of drug delivery systems.

To control the gastric residence time of the dosage forms in the fasted state, we need to identify factors important to gastric retention. Identification of such factors requires visualization of the behavior of dosage forms in the stomach. Of the many imaging techniques used in humans, gamma scintigraphy has become used widely to estimate the gastric emptying time of solid and liquid materials. In animals, the gastric emptying time is commonly estimated using gamma scintigraphy, radiography, or duodenal cannulation. These techniques, however, are self-limiting in that the response of the object of interest to dynamic events occuring in the stomach is difficult to visualize in real-time. The properties of the dosage form which are critical to gastric retention during fasted state activity can be more accurately assessed by visualizing the interactions between the dosage form and the gastric tissue undergoing peristalsis. Recently, enzyme-digestible hydrogels were developed as platforms for long-term oral drug delivery systems (28,29). These systems were studied in the canine stomach using ultrasound imaging (30). Ultrasound imaging provided real-time visualization of propulsion, retropulsion, and swelling of hydrogels in the stomach. More importantly, it was used to examine gastric tissue-gel interactions under varying gastric conditions. Ultrasound imaging, however, became difficult as more fluid emptied from the stomach. This was due to the attenuation of sound arising from gaseous particles in the lumen. Consequently, we combined ultrasound imaging with radiographic or fluoroscopic imaging. Using both imaging techniques, we were able to clearly observe the dynamic responses of an enzyme-digestible hydrogel in the fasted and fed stomach of a dog.

Our previous study in the fed state suggested that gastric retention would result when the propulsive efficiency of peristaltic contractions is minimized (30). In that study, the gastric retention of cylindrically shaped hydrogels (16 mm in length x 14 mm in diameter in the partially swollen state) was controlled by varying the degree of gastric distention using water. As the degree of gastric distention increased via repetitive administrations of water, neither the orad (retropulsive) nor the caudad (propulsive) movements of the gel were affected much by peristaltic contractions moving through the stomach. The degree of gastric tissue-gel interactions during peristalsis and thus the propulsive efficiency of peristaltic contractions was inversely related to the degree of gastric distention. Under maintained gastric distention, the gel primarily resided in the body of the stomach. As the degree of gastric distention was reduced due to the emptying of water, both the orad and caudad movements of the hydrogel were affected by peristaltic contractions moving through the stomach. Because of the intimate contacts between gastric tissues and hydrogel, the gel moved into the pyloric antrum. After all the water had emptied from the stomach, gel emptying resulted. This observation suggests that minimizing the propulsive efficiency of peristaltic contractions results in gastric retention. Minimizing the propulsive efficiency of peristaltic contractions in the fasted state may be achieved by manipulating the properties of the hydrogel. Therefore, we varied the hydrogel properties and examined how a hydrogel could be retained in the stomach under fasted conditions.

Materials and Methods

Albumin-crosslinked polyvinylpyrrolidone (PVP) hydrogels were prepared by free radical polymerization using 1-vinyl-2-pyrrolidinone (Aldrich) as a monomer, functionalized albumin (FA) as a crosslinking agent, and 2,2-azobis(2-methyl-propionitrile) (Eastman Kodak) (28). The monomer solution was degassed and purged with nitrogen followed by polymerization at 60°C for 18 h under nitrogen. Once polymerization was complete, the gels were removed and cut into cylinders. The prepared gels were purified and dried as described previously (29). The dried gels were then allowed to swell in a 4% (v/v) solution of diatrizoate meglumine/sodium diatrizoate (Gastrografin (GG), Squibb Diagnostics) for 32 h at 37°C. The GG-loaded gels were air dried for one week and later oven dried at 37°C for at least one week. Before each animal study, the dimension and weight of each dried gel were recorded. Under the present loading conditions, one gram of dried gel retained approximately 0.38 g of GG. Just prior to administration of the gel to a dog, the dried hydrogel was partially swollen in a 4% (v/v) GG solution for 15 min to impart lubricity on the gel surface to ensure safe transit to the stomach.

Imaging Technique. Standard lateral and ventrodorsal radiographs of the abdomen were made using a 1200 Ma, 150 KVP x-ray generator (General Electric Corporation, Milwaukee, WI). The animal was positioned in right lateral and dorsal recumbency. The same radiographic technique (76 & 86 KVP, 12 MaS), x-ray cassettes, film and film processing were used so that the decreasing opacity of the hydrogel could be monitored. Fluoroscopy was accomplished with the animal in sternal recumbancy. This allowed the hydrogel to be positioned in the body and/or pyloric antrum so that observations of the hydrogel movements could be made and recorded on video tape.

The ultrasound examination of the stomach was performed with a real time mechanical sector scanner (Interspec, Conshohocken, PA) using a 5 mHz transducer. After clipping the hair on the ventral right side of the body just caudal to the costal arch, copious amounts of an ultrasound transmission gel (Lafayette Pharmacal Inc., Lafeyette, IN) was applied to the skin. With the animal in a standing position, a transverse view of the cranial abdomen (sagittal view of the stomach) was obtained. The gastric tissue-gel interactions were observed and recorded on video tape. With a sagittal view of the hydrogel frozen on the monitor, the image of the gel was measured and recorded on a video format camera.

Animal Experiments. A healthy mongrel dog weighing 35 lb was used throughout the study. Before each experiment, the animal was fasted for at least 15 h and then radiographed just before the administration of a gel to ensure the absence of food in the stomach. For each experiment, 380 ml of water was instilled just before the administration of a gel. An additional 380 ml of water was given at every 30 min for up to 3 h when more exposure of the gel to water was desirable. Water was instilled using a stomach tube and syringe. To maintain fasted state conditions, no food or water was given to the animal after the last administration of water. Hydrogel integrity was varied by controlling the size of the glassy core in gels which achieved the same approximate dimensions after swelling. This was accomplished by using hydrogels with different dimensions in the dried state and by controlling hydrogel exposure time to water in the stomach. The initial size of the gel was varied from 18 mm in length by 26 mm in diameter to 29 mm in length by 29 mm in diameter. Smaller sized gels swelled to a completely amorphous network in about 4 h. Larger sized gels, however, contained a glassy core and an amorphous outer layer which could be assessed through ultrasound imaging.

Over the course of these experiments, if gastric retention in the fasted state was observed for 24 h, the animal was given a standard pelletized meal once-a-day. The purpose for this was to observe the effects of food on hydrogel retention and also to determine if the presence of the gel had any effect on inhibiting the transit of food.

Results.

Gastric retention of hydrogels with a partially swollen size of 18 mm in length by 26 mm in diameter was examined. After 2.5 h in the stomach, the gel increased in size to 29 mm in length by 33 mm in diameter. When the stomach lumen was largely distended, the hydrogel resided in the pyloric antrum where gel retropulsion was observed during peristaltic contractions. After 4 h, ultrasound detected that the gel was completely amorphous and that most of the water had emptied from the stomach. When a peristaltic contraction moved through the pyloric antrum, hydrogel deformation in response to a contraction was quite pronounced as seen through fluoroscopy. When a peristaltic contraction made initial contact with the hydrogel, the gel became compacted and partially deformed against the pyloric sphincter. As the contraction migrated along the gel's surface, the left half of the gel was compressed into the antrum while the right half was still compacted against the pyloric sphincter (Figure 1). The gel was subsequently retropelled once the contraction reached the distal end of the hydrogel. Fluoroscopic images taken after 5 h located the gel in the small intestine where the gel passed quite rapidly with fluid-like characteristics.

In another study, hydrogels with a partially swollen size of 29 mm in length by 29 mm in diameter were used. After 2.5 h in the stomach, a gel diameter of 36 mm was achieved. Ultrasound imaging at 3.5 h showed an acoustic shadow which was largely attributed to the glassy polymeric core. The presence of a glassy core had a profound influence on the gastric retention time. Once most of the fluids had emptied (approximately 4 h after gel administration), the gel resided in the pyloric antrum. When a peristaltic wave moved through the pyloric antrum, hydrogel deformation in response to contractions was limited to the amorphous edges of the hydrogel. When a peristaltic contraction made initial contact with the hydrogel, the gel became compacted against the pyloric sphincter. As the wave migrated axially toward the most distal region of the gel, deformation occurred only at the gel edge (arrow in Figure 2) before retropulsion. Over the first 24 h, under fasted conditions, the peristaltic contractions were ineffective in deforming the gel and propelling it through the pyloric sphincter. Thus, it appeared that minimizing gel deformation in response to peristaltic contractions resulted in gastric retention of the gel. Comparison with our previous study (*30*) suggests that both the gel size and the gel integrity are two parameters critical to hydrogel retention under fasted conditions. Once the water had emptied from the stomach, gel swelling is expected to be limited since the amount of gastric fluid under fasted conditions is not significant. Once the water had emptied from the stomach, however, assessment of the glassy core with ultrasound was no longer effective due to the attenuation of sound by gaseous paricles in the partially occluded lumen. In this case, fluoroscopic imaging was used to indirectly moniter the presence of the glassy core by assessing gel deformation in response to peristaltic contractions. During the 24 h-fasted conditions, gel deformation in response to peristaltic contractions gradually increased. The deformation, however, was only slightly greater than that observed in Figure 2.

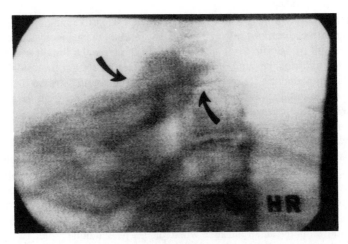

Figure 1. Fluoroscopic image illustrating maximum gel deformation in response to a single peristaltic contraction. When the peristaltic contraction passes over the gel, it becomes flattened (arrows) and subsequently retropelled. The image was taken 4 h after the administration of the gel.

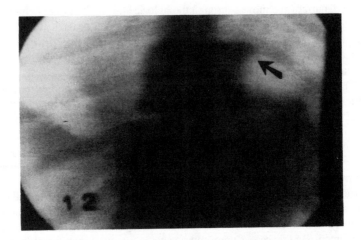

Figure 2. Fluoroscopic image illustrating minimum gel deformation in response to a single peristaltic contraction. When the peristaltic contraction passes over the gel, the deformation was limited to the edges of the gel (arrow). The image was taken 12 h after the administration of the gel.

On radiographic images taken 24 h after gel administration, noticeable traces of GG in the animal's colon were detected as indicated by arrowheads in Figure 3. It should be noted that the diffusional loss of GG from the gel was not significant. In vitro radiographic images made of a GG-loaded hydrogel in the simulated gastric fluid showed that the loss of GG from the gel was insignificant. This indicates that the release of GG under fasted conditions (Figure 3) was due to a mechanism other than simple diffusion. Over the 24 h of gastric retention of the gel, degradation on the gel surface became apparent as indicated by arrowheads in Figure 4. This suggests that the surface of the gel was attrited by the repeated contacts between the gel and the gastric tissue undergoing peristaltic contractions. Thus, it appears that surface erosion of the hydrogel was responsible for the release of GG from the gel. After the gel was retained in the stomach for 24 h under fasted conditions, the animal was given a standard pelletized meal at 24 h and 48 h. In the presence of food, the size of the gel reduced significantly. Figure 5 shows a gradual decrease in the hydrogel size over a period of 60 h. The size of the gel decreased much faster in the presence of food than in the fasted state. Food in the stomach apparently accelerated the surface erosion of the gel. The presence of the gel had no apparent effect on the emptying of food. Radiographic images made at 72 h illustrated that the hydrogel was neither in the stomach nor the intestines. This observation suggested that the presence of surface erosion combined with the effects of gel swelling and bulk degradation led to hydrogel disruption in the stomach.

Discussion

Our study indicated that the presence of a glassy core had a profound influence on gastric retention of hydrogels in the fasted state. When hydrogel deformation was minimal in response to gastric contractions, the contractions were ineffective in propelling the gel through the pyloric sphincter. It was this limited deformation in response to peristaltic activity that contributed to gastric retention. Although the hydrogels used in this study did not have exactly the same dimensions after the last instillation of water, their contrasting physical properties served as a useful comparison in understanding the effects of gel integrity on gastric retention. Integrity alone, however, does not guarantee hydrogel retention. In our previous studies, gels (16 mm in length by 14 mm in diameter) with only a moderate degree of swelling, emptied from the stomach within 2 h (*30*). The gel emptied from the stomach even though its deformation in response to gastric contractions was not significant. Thus, the size of the gel must also contribute to gastric retention. A third parameter that may influence gastric retention, under fasted conditions, is the gel's surface properties. The "slippery" surface of the gel arising from hydration and surface erosion appeared to facilitate both the movements of contractions along the gel surface and hydrogel retropulsion. Since only one type of hydrogel made of PVP was used in this study, the effects of gel surface properties on gastric retention requires further investigation. The important point here is that long-term gastric retention was achieved in the fasted state and was largely attributed to the combined effects of size,

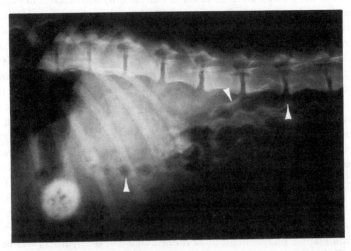

Figure 3. Lateral radiograph of the abdomen made at 24 h following administration of a GG-loaded gel. The presence of GG in the colon is indicated by arrowheads.

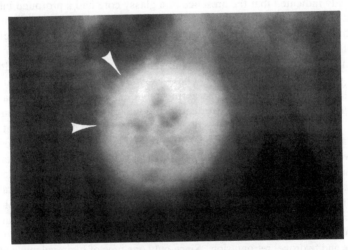

Figure 4. Close-up view of the GG-loaded gel in the pyloric antrum of the stomach. Surface erosion is indicated by arrowheads.

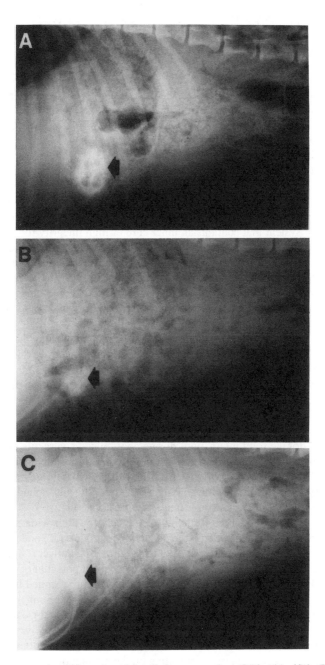

Figure 5. Lateral radiographs of the abdomen made at 36 h (A), 48 h (B), and 60 h (C). The size of a hydrogel (arrow) gradually reduced in the presence of food mainly due to the surface erosion.

integrity, and possibly surface lubricity. These effects served to minimize the propulsive efficiency of peristaltic contractions.

The surface erosion of gels in the stomach may be utilized as an effective means to deliver drugs to the body for extended periods of time. If we consider the GG to be a model drug, it is conceivable that sustained drug release can be achieved by a combination of erosion-controlled and diffusion-controlled release mechanisms. Although a more detailed analysis of drug release will be examined in future work, the preliminary findings from this study look quite promising with respect to once-a-day or even once-a-week drug delivery.

Even though the hydrogels used in this study may be useful in some veterinary applications (31), they are not yet ready for human studies since the initial size of the device is rather large for swallowing. The data collected in this present study, however, will be vital to the design of a new system which is more suitable for swallowing. Long-term gastric retention of hydrogels may be achieved by utilizing a biodegradable system that contains an inner core of high integrity for the minimum gel deformation and an outer core of high swelling capacity for the large gel size. This new design would be based on a 2-phase hydrogel network that contains a highly crosslinked, moderately swelling polymer as the inner phase and a loosely crosslinked, highly swellable polymer as the outer phase. In the stomach, this device would produce a size and integrity comparable to the present design, but its initial size would be small enough for easy swallowing.

In human studies, gamma scintigraphy has been used to study the GI transit of both digestible and nondigestible material because it is noninvasive and exposes the individual to only low radiation doses. One disadvantage of this approach is that the transit of only the labelled material can be monitored over time. When the contents of the stomach consist of nonlabelled digestible material and a labelled, nondigestible dosage form, this technique provides very little information regarding the gastric emptying of the dosage form relative to the digestible material (10). Furthermore, the gastric tissue-dosage form interactions during coordinated contractile events cannot be visualized. Thus, gamma scintigraphy does not provide a true understanding on the dosage form properties which are critical to controlling gastric retention. Consequently, when the drug delivery system is hydrophilic, ultrasound imaging may be utilized to study gastric retention in humans more accurately.

Hydrogels have received tremendous interest for their wide range of drug delivery applications (1). In particular, bioadhesive hydrogels (32) and pH-dependent swelling hydrogels (33) have been developed for oral drug delivery. Even though a great deal of research has been done to show the excellent potential of these systems in vitro, in vivo performance has either been untested or quite limited in success. As a result, in vivo studies are critical to the further development of these systems Because ultrasound imaging can monitor hydrogel swelling and gastric tissue-gel interactions in real-time, the utilization of hydrogels as oral drug delivery devices can be better assessed. Although there are some limitations with the ultrasound imaging technique such as poor resolution depending on the amount of subcutaneous fat and individual anatomical variations (34), this imaging technique warrents further consideration as a tool for the study and the rational design of long-term oral drug delivery systems.

In summary, the advantages of our approach to studying gastric retention are twofold. First, hydrogel response to the gastric environment can be noninvasively visualized in real-time. Second, a more efficient and justifiable approach to the development of long-term oral drug delivery systems can be achieved.

Acknowledgment

This study was supported in part by the ICI Pharmaceuticals Group.

Literature Cited

1. *Hydrogels in Medicine and Pharmacy;* Peppas, N.A., Ed.; CRC Press: Boca Raton, Florida, 1987; Vol. 3.
2. *Controlled Drug Delivery : Fundamentals and Applications;* Robinson, J.R.; Lee, V.H.L., Eds.; Marcel Dekker Inc.: New York, New York, 1987; 2nd edn.
3. Khosla, R.; Feely, L.C.; Davis, S.S. *Int. J. Pharm.* **1989**, *53,* 107.
4. Khosla, R.; Davis, S.S. *Int. J. Pharm.* **1989**, *52,* 1.
5. Davis, S.S.; Khosla, R.; Wilson, C.G.; Washington, N. *Int. J. Pharm.* **1987**, *35,* 253.
6. Mojaverian, P.; Ferguson, R.K.; Vlasses, P.H.; Rocci, M.L.; Oren, A.; Fix, J.A.; Caldwell, L.J.; Gardner, C. *Gastroenterology* **1985**, *89,* 392.
7. Davis, S.S.; Hardy, J.G.; Fara, J.W. *Gut* **1986**, *27,* 886.
8. Harris, D.; Fell, J.T.; Sharma, H.L.; Taylor, D.C. *J. Controlled Release* **1990**, *12,* 45.
9. Harris, D.; Fell, J.T.; Taylor, D.C.; Lynch, J.; Sharma, H.L. *J. Controlled Release* **1990**, *12,* 55.
10. Khosla, R.; Davis, S.S. *Int. J. Pharm.* **1990**, *62,* R9.
11. Davis, S.S.; Stockwell, A.F.; Taylor, M.J.; Hardy, J.G.; Whalley, D.R.; Wilson, C.G.; Bechgaard, H.; Christensen, F.N. *Pharm. Res.* **1986**, *3,* 4, 208.
12. Meyer, J.H.; Dressman, J.; Fink, A.; Amidon, G. *Gastroenterology* **1985** *89,* 805.
13. Park, H.M.; Chernish, S.M.; Rosenek, B.D.; Brunelle, R.L.; Hargrove, B.; Wellman, H.N. *Dig. Dis. Sci.* **1984**, *29,* 3, 207.
14. Minami, H.; McCallum, R.W. *Gastroenterology* **1984**, *86,* 1592.
15. Sirois, P.J.; Amidon, G.L.; Meye , J.H.; Doty, J.; Dressman, J.B. *Am. J.Physiol.* **1990**, *258,* G65.
16. Meyer, J.H.; Ohashi, H.; Jehn, D.; Thomson, J.B. *Gastroenterology* **1981**, *80,* 1489.
17. Meyer, J.H.; *Am. J. Physiol.* **1980**, *239,* G133.
18. Marvola, M.; Kannikoski, A.; Aito, H.; Nykanen, S. *Int. J. Phar.* **1989**, *53,* 145.
19. Hinder, H.A.; Kelly, K.A. *Am. J. Physiol.* **1977**, *233 (4),* E335.
20. Mroz, C.T.; Kelly, K.A. Surg. *Gynec. Obstet.* **1977**, *145,* 369.

21.   Houghton, L.A.; Read, N.W.; Heddle, R.; Maddern, G.J.; Downtown, J.; Toouli, J.; Dent, J. *Gastroenterology* **1988**, *94*, 1276.
22.   Houghton, L.A.; Read, N.W.; Heddle, R.; Horowitz, M.; Collins, P.J.; Chattern, B.; Dent, J. *Gastroenterology* **1988**, *94*, 1285.
23.   Code, C.F.; Marlett, J.A. *J. Physiol.* **1975**, *246*, 289.
24.   Szurszewski, J.H. *Am. J. Physiol.* **1969**, *217(6)*, 1757.
25.   Kelly, K.A.; Code, C.F. *Am. J. Physiol.* **1971**, *220*, 1, 112.
26.   Kelly, K.A.; Code, C.F.; Elveback, L.R. *Am. J. Physiol.* **1969**, *217*, 2, 461.
27.   DeWever, I.; Eeckhout, C.; VanTrappen, G.; Hellemans, J. *Am. J. Physiol.* **1978**, *4(6)*, **E661.**
28.   Park, K. *Biomaterials* **1988**, *9*, 435.
29.   Shalaby, S.W.S.; Park, K. *Pharm. Res.* **1990**, *7*, 8, 816.
30.   Shalaby, S.W.S.; Park, K.; Blevins, W.E. *Proceed. Intern. Symp. Control.Rel. Bioact. Mater.* **1990**, *17*, 132.
31.   Cardinal, J.R. *J. Controlled Release* **1985**, *2*, 393.
32.   Park, K.; Robinson, J.R. *Int. J. Pharm.* **1984**, *19*, 107.
33.   Brannon-Peppas, L.; Peppas, N.A. *J. Controlled Release* **1989**, *8*, 267.
34.   King, P.M.; Adam, R.D.; Pryde, A.; McDicken, W.N.; Heading, R.C. *Gut* **1984**, *25*, 1384.

RECEIVED March 19, 1991

LIPOSOMAL DRUG DELIVERY

Liposomal Drug Delivery

Although many of the materials used to prepare liposomes are not polymers, these amphiphilic materials do form very ordered structures with specific orientations similar to many polymers. More recently, polymers have been used to stabilize liposomes either by a coating of the liposome or using materials which can polymerize after the liposome has been formed. The four chapters in this section describe the preparation of liposomes with unusual characteristics and the potential for more stability and direct targeting to the site of treatment.

In the first chapter of this section (Chapter 22), the formation of liposomes from a liquid crystalline phase known as the cubic phase is described. Monoolein, a common food additive, is used to prepare these liposomes and the release of a number of oligopeptide drugs is discussed. In Chapter 23, the formation of lipid microspheres which resemble liposomes in their tissue distribution is presented. The clinical efficacy of these lipid emulsions in treating a number of cardiovascular diseases appears significant. In addition, the conjugation of a model peptide to a lipid material, lecithin, is shown to give tissue distributions very similar to those of the lipid microspheres. The next paper, Chapter 24, presents the efforts of research aimed at targeting liposomes to a specific organ within the body. The preparation, optimization, and drug delivery potential of these organ-specific immunoliposomes are shown. The final paper in this section, Chapter 25, describes the preparation of polymeric liposomes with biodegradable linkages. The polymerization of the lipid improves the stability of the synthetic liposomes and the hydrolyzable linkage between the lipid and the polymer chain allows for the generation in the body of a water-soluble polymer. It is hoped that further research of this type will be conducted so that liposomal delivery systems will receive greater usage for drug delivery.

Chapter 22

Cubic Phases as Delivery Systems for Peptide Drugs

B. Ericsson[1,5], P. O. Eriksson[2], J. E. Löfroth[3], and S. Engström[4,6]

[1]Ferring AB, POB 30561, S–200 62 Malmö, Sweden
[2]Physical Chemistry, Umeå University, S–901 87 Umeå, Sweden
[3]Drug Delivery Research, AB Hässle, S–431 83 Mölndal, Sweden
[4]Food Technology, Chemical Center, POB 124, S–221 00 Lund, Sweden

Monoolein spontaneously forms a cubic liquid crystalline phase with water which can co-exist with excess water. Its amphiphilic character and highly ordered structure makes the cubic phase an interesting candidate for drug delivery. We have studied the interaction of some oligopeptide drugs - Desmopressin, Lysine Vasopressin, Somatostatin and the Renin inhibitor (H214/03) - with the cubic phase by means of phase behaviour, self-diffusion and in vitro and in vivo release. In addition, the hindered enzymatic degradation of a peptide drug in the cubic phase is being investigated as well. The results show that oligopeptides can be solubilized in the cubic phase in an amount of about 5-10% (w/w). The self-diffusion data indicate that Desmopressin interacts significantly with the monoolein-water interface. The in vitro release data show a time-dependence (*i.e.* $t^{1/2}$) as is expected of a matrix system. The in vivo release data from intramuscular and subcutaneous depots of cubic phase containing Desmopressin and Somatostatin, show a constant Desmopressin-like and Somatostatin-like immunoreactivity over six hours. Experiments show that peptide drugs in the cubic phase are protected against enzymatic cleavage in simulated intestinal fluid.

Polar lipids form different kinds of aggregates in water, which in turn give rise to several phases, such as micellar and liquid crystalline phases. Among the latter, the lamellar phase (L_α) has received the far greatest attention from a pharmaceutical point of view. The lamellar phase is the origin of liposomes and helps in stabilizing oil-in-water (O/W) emulsions. The lamellar structure has also been utilized in creams. We have focused our interest on another type of liquid crystalline phase - the cubic phase

[5]Current address: Kabi Invent AB, Blickagången 6D, S–141 52 Huddinge, Sweden
[6]Corresponding author.

0097–6156/91/0469–0251$06.00/0
© 1991 American Chemical Society

- and in particular the cubic phase formed of monoolein and water. Monoolein is a common food additive and has also been frequently used in cosmetics.

Cubic Phases in the Monoolein-Water System. Monoolein (or glyceryl monooleate, see Table I) is a polar lipid which swells in water, giving rise to a reversed micellar phase (L_2) and three types of liquid crystalline phases (lamellar, reversed hexagonal and the cubic phase) (*1*). Its phase behaviour in water is given in Figure 1. A large cubic phase region dominates the phase diagram (in fact the region consists of two cubic phases, G and D, of similar structures). An interesting property of the cubic phase D is its ability to co-exist with excess water (the monoolein solubility in water is about 10^{-6} M).

The structure of the cubic phase, which is outlined in Figure 1, consists of a curved bilayer extending in three dimensions, separating two congruent networks of water channels. The water pore diameter of the fully swelled phase is about 5 nm (*1*), and the phase is very viscous. Taken together, these properties add up to make the monoolein-water cubic phase an in situ forming biodegradable matrix system. Its biodegradability comes from the fact that monoolein is subject to lipolysis due to different kinds of esterase activity in different tissues.

Monoolein will also form the cubic phase together with lecithin (e.g. dioleoyl phosphatidylcholine, see Figure 1), but above about 50% (w/w) lecithin the cubic phase is transformed into the lamellar phase (*2*). Moreover, water may be replaced by glycerol, completely or partly, in the cubic phase. Vegetable oils, e.g. sesame oil, can be incorporated to some extent (a few percent) in the cubic phase, and the same holds for bile salts.

Drugs in the Cubic Phase. In order to function as a drug delivery system, the monoolein-water cubic phase has to be able to dissolve or disperse drugs. In this respect the cubic phase reveals a great flexibility, since drugs of very different polarity and size may be incorporated (*3*). Of special interest is its ability to solubilize peptides and proteins. The reason for the flexibility of the system is its amphiphilic nature, since the cubic phase roughly consists of equal amounts of lipid and water. The bilayer structure gives rise to a large interfacial area (≈ 400 m^2/g cubic phase), which should promote the incorporation of amphiphilic substances, e.g. peptides and proteins.

In this work we will focus on the use of the cubic phase as a delivery system for oligopeptides - Desmopressin, Lysine Vasopressin, Somatostatin and the Renin inhibitor H214/03. The amino acid sequences of these peptides are given in Table I. The work focuses on the cubic phase as a subcutaneous or intramuscular depot for extended release of peptide drugs, and as a vehicle for peptide uptake in the GI-tract. Several examples of how the peptide drugs interact with this lipid-water system will be given in terms of phase behaviour, peptide self-diffusion, in vitro and in vivo release kinetics, and the ability of the cubic phase to protect peptides from enzymatic degradation in vitro. Part of this work has been described elsewhere (4-6).

Table I. Molecular formulas

Monoolein	CH_3-$(CH_2)_7$-CH=CH-$(CH_2)_7$-CO-O-CH_2-CH(OH)-CH_2-OH
Lecithin (*e. g.* DOPC)	di-oleoyl-phosphatidylcholine
Desmopressin (dDAVP)	S-CH_2-CH_2-CO-Tyr-Phe-Gln-Asn-Cys-Pro-D-Arg-Gly-NH_2
Renin inhibitor (H214/03)	Boc-Phe-His-Leu-[OH]-Val-Ile-His-OH
Lypressin (LVP)	Cys-Tyr-Phe-Gln-Asn-Cys-Pro-Lys-Gly-NH_2
Somatostatin (SRIF)	Ala-Gly-Cys-Lys-Asn-Phe-Phe-Trp-Lys-Thr-Phe-Thr-Ser-Cys

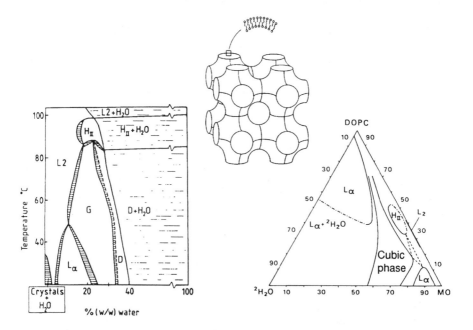

Figure 1. Phase diagrams of the systems MO-H_2O (Adapted from ref. 1) and MO-DOPC-H_2O (at 28°C, Adapted from ref. 2), respectively. G (gyroid) and D (diamond) are cubic phases with similar structures to the type shown schematically in the figure (Adapted from ref. *18*).

Materials and Methods

Materials. Monoolein (MO) (98% monoglyceride, 91% monoolein, Grindsted A/S, Denmark), soy bean lecithin (LE) (Epicuron 200, Lucas Meyer, Germany), Na taurocholate (STC) (Sigma T4009), sucrose (S) (Sigma S8501), chymotrypsin (49 units/mg, Sigma C4629), Tris buffer (Sigma T3253, T1503), deuterium oxide (^2H_2O) (Ciba-Geigy, Schweiz) and acetonitrile (Merck) were used as purchased. The peptides Desmopressin (dDAVP, 1-desamino-8-D-arginine vasopressin), Lysine vasopressin (LVP) and synthetic Somatostatin (SRIF)were obtained from Ferring AB (Malmö Sweden), and the Renin inhibitor (H214/03) was obtained from AB Hässle. The peptides were used without further purification. Distilled water was of Milli-Q quality (Millipore).

Sample Preparation. Liquid crystalline phases, i.e. cubic and lamellar phases, were prepared by weighing the components in stoppered test tubes or into glass ampoules (which were flame-sealed). Water soluble substances were added to the system as water solutions. The hydrophobic substances were dissolved in ethanol together with MO, and the ethanol was then removed under reduced pressure. The mixing of water and MO solutions were made at about 40°C, by adding the MO solution dropwise. The samples for the in vivo study were made under aseptic conditions. The tubes and ampoules were allowed to equilibrate for typically five days in the dark at room temperature. The phases formed were examined by visual inspection using crossed polarizers. The compositions for all the samples used in this work are given in Tables II and III.

Self-Diffusion of Desmopressin and Monoolein by NMR. The self-diffusion coefficient was measured with the NMR diffusion technique using a Bruker MSL 100 spectrometer. Two magnetic field gradient pulses were applied at either side of the 180-degree pulse in a $[90_x$-τ-180_y-τ-echo] spin echo sequence (7,8) (Figure 2). Due to diffusion, the amplitude of a component in the spin-echo spectrum is attenuated according to (7)

$$A = A_0 \exp[-\gamma^2 g^2 \delta^2 (\Delta - \delta/3) D] \qquad (1)$$

where γ is the magnetogyric ratio of the nucleus and Δ is the time between the gradient pulses. The diffusion coefficient, D, was obtained from a two parameter (A_0 and D) non-linear least-squares fit of equation (1) to the peak amplitude as a function of δ, the duration of the gradient pulses. The magnitude of the gradient pulses, g, was determined by measurements on distilled water for which reliable self-diffusion data exists.

In vitro Release of Desmopressin and Lysine Vasopressin. The release experiments were carried out in triplicate according to the USP paddle method (100 rpm, 37°C, 500 mL water). Appropriate amounts of the cubic phase (typically 100 mg) were weighed in small cups (depth=2 mm, diameter=8 mm) drilled non-

Table II. Composition of systems used in the NMR and release experiments, and diffusion coefficients, D, of peptides (PE)

Peptide	Composition, %w/w				$D \cdot 10^{12}$, $m^2 s^{-1}$	Figure:Curve
	PE	MO	LE	H_2O		
NMR self-diffusion at 25 °C						
dDAVP	5	-	-	95[a]	225 ± 10	
NMR self-diffusion at 40 °C						
dDAVP	0	60	-	40[a]	32 ± 4[b]	
	10	50	-	40[a]	24 ± 4	
					31 ± 4[b]	
In vitro release at 37 °C						
dDAVP	4	68	-	28	18.7	3:dDAVP-0
	4	55	25	16	9.4	
	4	45	40	11	7.0	
	4	65	10[c]	21	30.2	3:dDAVP-S-10
	4	65	2[d]	29	23.0	
LVP	4	68	-	28	13.3	3:LVP-0
	4	55	25	16	4.8	3:LVP-LE-25
	4	45	40	11	0.34	3:LVP-LE-40
	4	63	11[c]	22	7.7	
	4	65	2[d]	29	12.0	
H214/03	4	65	-	31	≈0[e]	
In vivo release in the rabbit						
dDAVP	1.3	90	-	8.7	-	4a:dDAVP-sc
	1.3	90	-	8.7	-	4a:dDAVP-im
SRIF	2	58	-	40	-	4b:SRIF-0
	2	53	13	32	-	4b:SRIF-LE-13
	2	42	28	28	-	4b:SRIF-LE-28

[a] ^2H_2O

[b] Diffusion coefficient of monoolein

[c] Sucrose (S)

[d] Sodium taurocholate (STC)

[e] Not measurable in the release experiments

Table III. Composition of mixtures used in the enzymatic degradation experiments

System	Conc	Comments	k^a/min^{-1}	Relative rate	Figure:Curve
H214/03 in buffer+ enzyme/buffer	10 µg/ml 100 µg/ml	homogeneous solution	0.07967	100[b]	5:H214-sol
H214/03/MO+ enzyme/buffer	10 µg/80 mg 100 µg/ml	cubic phase formed	0.00546	5.7	5:H214-cub
H214/03/MO+ enzyme/buffer + STC 5 mM 10 mM 30 mM	10 µg/80 mg 100 µg/ml	dispersion of cubic phase	0.00451 0.00562 0.01087	5.7 7.1 13.6	5:H214-STC-5 5:H214-STC-10 5:H214-STC-30
LVP in buffer+ enzyme/buffer	20 µg/ml 10 µg/ml	homogeneous solution	0.01833	100[b]	
LVP/MO+ enzyme/buffer	20 µg/80 mg 10 µg/ml	cubic phase formed	0.00603	32.9	

[a] The pseudo-first order rate constant.
[b] The rates in the homogeneous solutions were both assigned a relative rate of 100, since they were not comparable with each other as discussed in the text.

centrically in cylindrical Teflon discs (height=15 mm, diameter=36 mm). Care was taken to get the exposed area of the phase as even as possible with the help of micro slides. The paddles were placed 20 mm above the Teflon discs. Released amounts of the peptides were determined with HPLC (dDAVP: 75/25% v/v phosphate buffer/CH_3CN, pH=3.1; LVP: 88/12% v/v phosphate buffer/CH_3CN, pH=3.1). The samples were examined in a polarizing microscope before and after the experiments, in order to reveal any possible texture changes.

In vivo Release of Desmopressin and Somatostatin. The in vivo release of Desmopressin and Somatostatin after subcutaneous and intramuscular injections of the peptide in the cubic or the lamellar phase has been studied in the rabbit. Blood was sampled at regular intervals, and systemically absorbed Desmopressin and Somatostatin were determined as the specific immunoreactitvity in plasma of the actual peptide. For details of the analyses with dDAVP, consult ref. 9. For comparison, Desmopressin-like and Somatostatin-like immunoreactitvity (dDAVP-LI and SRIF-LI) in plasma after intravenous bolus injections of the two peptides were determined as well.

The cubic phases with Somatostatin were allowed to swell to the water swelling limit, while the Desmopressin preparations were of the lamellar phase type, i.e. with low water content, and were therefore assumed to swell in vivo. Approximately 0.5 g of either cubic or lamellar phase was injected, corresponding to 2.5-3.0 mg peptide per kg bodyweight.

Enzymatic Degradation of H214/03. The degradation experiments were carried out with molecular mixtures of H214/03 or LVP and MO. These were obtained from mixtures of the peptide and MO dissolved in 99.5% ethanol, from which the ethanol was evaporated under reduced pressure at room temperature for about two days. Appropriate amounts of melted mixtures were then weighed in small Erlenmeyer flasks. Care was taken to insure that the bottom of each flask was covered with the mixture, which was then left to freeze. In this way the area and the thickness of the lipid/peptide layer was comparable in the flasks. Typically five flasks were used in each experiment.

The reaction was started by transferring 1 mL of the enzyme/buffer/bile salt solution (pH=7.2, 37°C) to each flask placed in a thermostated shaker at 37°C. Experiments were carried out without lipid and bile salt as well, and in these experiments equal amounts of stock solutions of the enzyme in buffer and peptide in buffer were mixed in the flasks at time zero, to give the indicated concentrations (see Table III). The reactions in the flasks were stopped by adding 0.5 ml acetonitrile at different times. The total amount of intact peptide remaining in a flask was determined by HPLC, after the content was dissolved by adding ethanol.

Results and Discussion

Oligopeptides in the Monoolein-Water System. Most oligopeptides are amphiphilic molecules, i.e. they have both hydrophilic and lipophilic characteristics, and are therefore expected to interact with the large interface of the cubic phase. It was found that the oligopeptides investigated in this work could be incorporated to 5-10% (w/w) in the monoolein-water system, while still showing a cubic phase structure as revealed by the polarizing microscope. Above certain concentrations of the water-soluble oligopeptides, the cubic phase was typically transformed to the lamellar phase, probably due to electrostatic repulsion at the monoolein-water interface caused by the peptides.

NMR Self-Diffusion of Desmopressin. The NMR-diffusion technique (*3,10*) offers a convenient way to measure the translational self-diffusion coefficient of molecules in solution and in isotropic liquid crystalline phases. The technique is nonperturbing, in that it does not require the addition of foreign probe molecules or the creation of a concentration-gradient in the sample; it is direct in that it does not involve any model dependent assumptions. Obstruction by objects much smaller than the molecular root-mean-square displacement during Δ (approx 1 μm), lead to a reduced apparent diffusion coefficient in equation (1) (*10*). Thus, the NMR-diffusion technique offers a fruitful way to study molecular interactions in liquids (*11*) and the phase structure of liquid crystalline phases (*11,12*).

The result of a typical diffusion measurement is shown in Figure 2. In the ^{1}H-NMR spectrum of a cubic phase of monoolein and ^{2}H$_2$O with 10% Desmopressin, the signals from the aromatic residues (Tyr and Phe) in Desmopressin, appear in a spectral region which does not contain any signals from the lipid. Therefore, the peptide and lipid diffusion coefficients could be determined separately (Table II), and in Figure 2 the spectra from such an experiment are shown. The lipid diffusion coefficient was also determined in a cubic phase in the absence of Desmopressin.

The Desmopressin diffusion coefficient in the cubic phase at 40°C (D=0.24 x 10^{-10} m^{2}s^{-1}) is about a factor 9 smaller than in ^{2}H$_2$O-solution at 25°C (D=2.25 x 10^{-10} m^{2}s^{-1}), a difference which is larger than what is expected from pure obstruction effects; a reduction factor of three is expected from the inclusion of a solute in the water channels of the cubic phase (*13*). Thus, the results indicate an interaction between the peptide and the lipid matrix and/or membrane surface, especially since the peptide and lipid diffusion coefficients are very similar in the cubic phase (Table II).

In Vitro Release of Desmopressin and Lysine Vasopressin. One advantage with the monoolein cubic phase, in the context of drug release, is its ability to co-exist with excess water. This fact, together with the stiffness of the phase, makes a release experiment simple, since one does not need any support to keep the system in place. The phases used in the experiments were isotropic both before and after the experiments as judged by inspection in the polarizing microscope. With respect to the original sample composition, the phases could therefore be regarded as cubic. The results from the release experiments are presented in Table II, and some typical release profiles are shown in Figure 3. The full drawn curves are the best fit of equation 2 to the data for times less than 400 minutes, where M(t) is the released

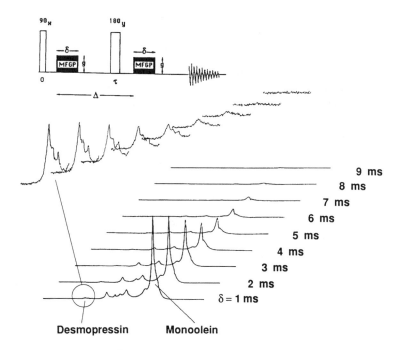

Figure 2. Spin-echo ^{1}H-NMR spectra from a diffusion experiment with a cubic phase of dDAVP (10%), MO (60%) and ^{2}H$_2$O (40%). Temperature 40°C, t=20 ms, Δ=24 ms, g=119 gauss/cm and δ=1.0, 2.0..., 9.0 ms. The inset shows the aromatic region originating from dDAV P at a higher amplification. Also shown is the pulse sequence used in the NMR-diffusion method (see text for details).

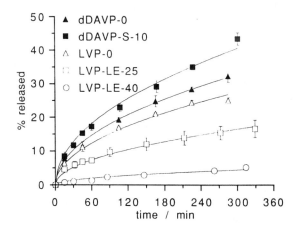

Figure 3. In vitro release experiments for dDAVP and LVP in various cubic phases at 37°C. The compositions of the samples are listed in Table II.

$$M(t) = 2 M(\infty) \sqrt{\frac{Dt}{h^2 \pi}} \qquad (2)$$

amount at time t, M (∞) the total amount of drug in the sample, D the mutual diffusion coefficient and h the depth of the release cup (2 mm). Equation 2 assumes perfect sink conditions, and that the diffusion in the phase was one dimensional over the depth of the cup, and that no significant stagnant water diffusion layer exists. Moreover, equation 2 is an approximation for "early times" as described elsewhere (14). Values of D obtained from the linear least-squares fits for the different experiments are summarized in Table II.

Although equation 2 is an approximation, the results show that the release mechanism was the same during essentially the whole release experiment (24 hours), although initially a small amount of water swelling most certainly occurs (swelling to the maximum water content, which is about 35% (w/w) for this monoolein sample). According to the applied model the rate determining step was the diffusion of the peptide in the phase. Experiments dDAVP-0 and LVP-0 in Figure 3 show that LVP was more retarded in the monoolein cubic phase than dDAVP. However, the partition coefficients of dDAVP and LVP between a cubic phase of MO/H_2O and H_2O had been determined in preliminary experiments to be both near zero. (The experiments were carried out by determining the peptide concentration in equilibrium with MO and water. Also the solubility of dDAVP and LVP in water are both more than 15% (w/w) as determined from solubility studies.) Thus, the release was expected to be 100%. Also, since both peptides are in their solubilized states in the phases, as judged from phase studies, different dissolution rates of the peptides could not explain the different rates of the peptides. LVP is slightly more hydrophobic than dDAVP, as determined by HPLC retention times. Thus, the results were indicative of different interactions between MO and the two peptides.

Tentatively, the interaction between the peptides and the polar parts of the lipid was an important rate determining factor. This could also explain the results obtained when up to 40% LE was incorporated in the cubic phases of dDAVP and LVP. In these experiments the release rates decreased with increasing amount of LE, the effect being more pronounced for LVP than for dDAVP. These retardations could, however, also be explained by the lower amount of water in the fully swelled cubic phases with lecithin according to the phase diagram of $MO/LE/H_2O$ shown in Figure 1, thus creating a larger obstruction effect to the diffusion in the water channels of the phase. Increased release rates were obtained for dDAVP when 10% (w/w) sucrose and 2% (w/w) sodium taurocholate (a bile salt) was incorporated in the cubic phase. The mechanisms behind these enhancements have not been further examined. Moreover, the release of H214/03 from the cubic phase is essentially zero, due to the low solubility of this peptide drug in water, i.e. H214/03 is located in the lipid domain of the cubic phase.

Finally, it is interesting to note the close agreement between the diffusion coefficient of Desmopressin obtained by NMR on the one hand, with that obtained from the release experiments on the other. This agreement gives further support to the conclusion that the release is caused by diffusion of the peptide in the cubic phase. The NMR technique offers a method to determine the diffusion coefficient within the phase, which then can be inserted into more complicated release models where other mechanisms may occur.

In Vivo Release of Desmopressin and Somatostatin in the Rabbit. These studies were undertaken in order to investigate the ability of the cubic phase to act as a subcutaneous (sc) or an intramuscular (im) depot of peptide drugs.

Desmopressin. Desmopressin is an analogue to the endogenous antidiuretic peptide hormone Vasopressin in which the modifications of the N-terminous amino acid and the replacement of the L-Arg for a D-Arg in position 8, significantly increases its biological stability. In this investigation, the half-life of dDAVP in the rabbit after intravenous administration was determined to be approximately 45 minutes.

The dDAVP preparations used in this study were prepared in low water contents so that the lamellar phase was formed, which in turn was injected into the rabbits. The reason for this was the fact that the lamellar phase with its mucous-like rheology is easier to inject than the stiff cubic phase. Since the lamellar phase swells into the cubic phase in excess water according to the phase diagram in Figure 1, a phase transition was expected also in the in vivo situation. The transition was found to be very fast as judged by inspection of the injection site immediately after administration.

As can be seen in Figure 4a, incorporation of dDAVP in the liquid crystalline phase significantly prolongs the apparent half-life of the peptide. No decline in plasma dDAVP-LI was found during the observation period of five hours, and the level of dDAVP-LI in plasma thus seems to correlate with an apparent zero-order release process of dDAVP. No difference in plasma dDAVP-LI could be found between sc and im administration.

Somatostatin. Somatostatin is an endogenous peptide hormone involved in e.g. the control of the release of Somatomedin, Insulin and Pancreatin. Due to its biological role, Somatostatin has a very low biological stability. The half-life in the rabbit after intravenous administration has been determined to approximately 90 seconds in this investigation. After sc or im administration, the apparent half-life is somewhat longer, close to 10 minutes, probably due to the absorption of the peptide from the injection site into the systemic circulation.

The inclusion of Somatostatin into a cubic phase markedly prolonged the apparent half-life. As can be seen in Figure 4b, the plasma level of Somatostatin-like immunoreactivity (SRIF-LI) after sc administration of cubic preparations of SRIF was nearly constant during the observation time of six hours, and no decline in the plasma level could be seen. During this period, the level of SRIF-LI in plasma thus seems to correlate with a zero-order release process of SRIF (cf. dDAVP). In Figure 4b is also shown the plasma SRIF-LI after sc administration of the peptide solubilized in cubic phases with varying ratios of MO/LE. It can be seen that increasing the amount of LE into the cubic phase decreases the release rate in accordance with the in vitro results on dDAVP and also on SRIF (6).

Tissue Response to Monoolein Phases. During the short term experiments (six hours) with the LC phases reported above, no tissue reactions were found. In order to investigate the effect of prolonged exposition of the tissue to LC phases, rabbits were injected subcutaneously and intramuscularly with either lamellar or cubic phases. After five days the animals were killed, and the injection sites were carefully examined.

In all cases a connective tissue encapsulation of the injected phases was found, and, in most cases, some of the phase remained. In a few animals there were signs of irritation, either in the connective tissue capsule or in the surrounding tissue. Moreover, injections of the lamellar phase, which swell to the cubic phase in vivo, seemed to be slightly more irritating than injections of the fully swelled cubic phase, most probably due, in the former case, to dehydration of the surrounding tissue. However, no difference was found between sc and im administration. It should be noted that the monoolein used was not of pharmaceutical grade.

Enzymatic Degradation of H214/03 in Simulated Intestinal Fluid. Monoolein is a metabolite during fat digestion, and there is strong evidence that the cubic phase is formed in vivo during this process (15). We therefore studied the ability of the cubic phase to protect an oligopeptide from enzymatic cleavage by the enzymes in the GI-tract. Moreover, since bile salts, such as sodium taurocholate (STC), form mixed micelles with monoolein and thus dissolve the cubic phase (16), we made some experiments in the presence of various concentrations of STC.

The results from the enzymatic degradation experiments are given in Table III and Figure 5. The reactions could be described by pseudo-first order reaction kinetics. However, the concentrations of the peptides and the enzymes were not the same in the experiments with the different peptides. Thus, the rates between the peptides cannot directly be compared, as is indicated in Table III. The peptide H214/03 is a hydrophobic substance, and it was found from the release experiments (see Table II) that no detectable amounts of peptide had been released from a cubic phase of MO/H_2O during 24 hours. This fact was reflected in the degradation experiments. The incorporation of H214/03 in the cubic phase protected the peptide from degradation, the rate decreasing to 5.7% of the rate in a homogeneous solution. The degree of protection of LVP was lower and gave a relative rate with the cubic phase of 32.9%. Thus, the degradation rates were mainly dependent on the release rates of the peptide out from the phase to the surface during the experiment. The diffusion of the chymotrypsin enzyme into the cubic phase was most probably negligible in these cases due to its large size compared to the water pores in the cubic phase.

The influence of STC on the degradation rates were examined at different STC concentrations, 5, 10 and 30 mM. The concentrations were chosen to reflect the intestinal concentrations expected in normal human subjects. According to the phase diagram STC/MO/^2H_2O by Svärd et al (16), the MO/H_2O phase at 5 mM bile salt

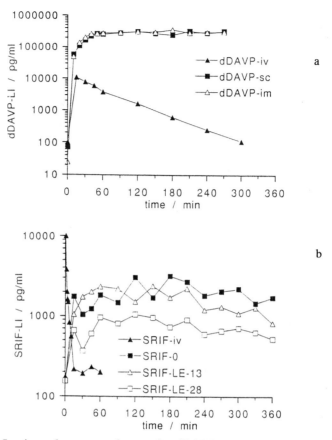

Figure 4. In vivo release experiments for dDAVP (4a) and SRIF (4b) in various cubic phases injected subcutaneously and intramuscularly in a rabbit model. The compositions of the samples are listed in Table II.

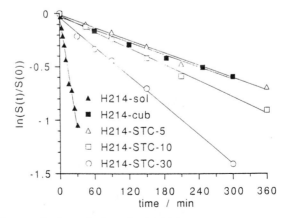

Figure 5. Enzymatic degradation of H214/03 in simulated intestinal fluid. $S(t)/S(0)$ is the fraction of remaining peptide at time t. The compositions of the samples are listed in Table III.

results in a dispersed cubic phase. At 10 mM it consists of lamellar and cubic structures, and with 30 mM bile salt can be regarded as being mainly lamellar. It was found that despite the high degree of destruction of the cubic phase with 30 mM bile salt, the peptide H214/03 was still protected against enzymatic degradation. Similar results had been obtained when the cubic phase was preformed, before addition of the enzyme solution.

Experiments in homogeneous solutions without MO but with added bile salt did not show any significant changes in the degradation rate as compared to the experiment without bile salt. Thus, we attribute the decrease of the degradation rates of the peptides to protective effects emanating from the liquid crystalline phases, and not to decreased enzyme activity in the presence of bile salts. The degradation rate would thus be dependent on the destruction rate of the cubic phase by the bile salt. The more hydrophilic peptide LVP would tentatively be better protected with a modified cubic phase containing e.g. lecithin. As was shown in the release rate experiments (see Table II), the presence of 40% LE decreased the diffusion coefficient to around 3% of that found without lecithin, which would imply an important protection factor.

Conclusion

It should be pointed out that cubic phases, such as the one discussed in this work, frequently occur in lipid-water systems (17), and the concept of using cubic phases as drug vehicles is therefore not limited to the use of monoolein only. From a toxicological stand-point, it is tempting to try to use membrane lipids, such as phospholipids, instead of monoolein for parenteral depot preparations (18-20).

Acknowledgments

We are indebted to Richard Andersson, Annike Andreasson, Pontus Bergstrand, Gunilla Hanisch, Karin Hansson, Stefan Leander, Katarina Lindell and Mats Olin for fruitful discussions and technical assistance.

Literature Cited

1. Hyde, S.T., Andersson, S., Ericsson, B. and Larsson, K. Z. Kristallogr. 1984, 168, 213.
2. Gutman, H., Arvidsson, G., Fontell, K., Lindblom, G. In Surfactants in Solution; Mittal, K.L., Lindman, B., Eds., Plenum Press, New York, 1984, Vol. 1, pp 143-152.
3. Engström, S. Lipid Technology, 1990, 2, 42.
4. Engström, S., Larsson, K. and Lindman, B., In Proceed. Intern. Symp. Control. Rel. Bioact. Mater., 15, Controlled Release Society, Inc., 1988, pp. 105-106.
5. Löfroth, J.-E., Andreasson, A. and Rehmberg, G., In Proceed. Intern. Symp. Control. Rel. Bioact. Mater., 15, Controlled Release Society, Inc., 1988, pp. 380-381.

6. Ericsson, B., Leander, S. and Ohlin, M., In *Proceed. Intern. Symp. Control. Rel. Bioact. Mater., 15,* Controlled Release Society, Inc., 1988, pp. 382-383.
7. Stejskal, E. O., Tanner, J. E. *J. Chem. Phys.* **1965,** *42,* 288.
8. Stilbs, P. *Progr. Nucl. Magn. Reson. Spectrosc.* **1986,** *19,* 1.
9. Harris, A.S., Ohlin, M., Lethagen, S., Nilsson, I.-M. *J. Pharm. Sci.* **1988,** *77,* 337.
10. Jönsson, B., Wennerström, H., Nilsson, P. G., Linse, P. *Coll. Polym. Sci.* **1986,** *264,* 77.
11. Eriksson, P. O.; Lindblom, G.; Burnell, E. E.; Tiddy, G. J. T. *J. Chem. Soc. Faraday Trans.* **1988,** *84,* 3129.
12. Eriksson, P. O.; Lindbom, G.; Arvidson, G. *J. Phys. Chem.* **1987,** *91,* 846.
13. Rilfors, L.; Eriksson, P. O.; Arvidson, G.; Lindblom, G. *Biochemistry* **1986,** *25,* 7702.
14. Cardinal, J.R., In *Medical Applications of Controlled Release;* Langer, R.S.; Wise, D.L., Eds.: CRC Press: Boca Raton, Fl., 1984, Vol. 1; pp 41-67.
15. Patton, J.S. and Carey, M.C. *Science* **1979,** *204,* 145.
16. Svärd, M., Schurtenberger, P., Fontell, K., Jönsson, B. and Lindman, B. *J. Phys. Chem.* **1988,** *92,* 2261.
17. Fontell, K. *Colloid Polym. Sci.* **1990,** *268,* 264.
18. Lindblom, G. and Rilfors, L. *Biochim. Biophys. Acta* **1989,** *988,* 221.
19. Larsson, K., Cubic lipid-water phases: Structures and biomembrane aspects. *J. Phys. Chem.* **1989,** *93,* 7304.
20. Seddon, J.M. *Biochim. Biophys. Acta* **1990,** *1031,* 1.

RECEIVED March 19, 1991

Chapter 23

Lipid Microspheres and Lecithinized-Polymer Drug Delivery Systems

Y. Mizushima and R. Igarashi

Institute of Medical Science, St. Marianna University, Kawasaki 216, Japan

Prostaglandins and some bioactive polypeptides are autacoids which are produced and act locally in the body as needed. The use of drug delivery systems in the treatment of diseases by these drugs may become necessary to achieve the needed pharmacokinetics. Lipid microspheres with an average diameter of 0.2μ resemble liposomes in terms of tissue distribution. They accumulate in the reticuloendothelial organs, inflamed tissues and particularly vascular lesions. In our studies, we have determined the distribution of the lipid microspheres in the vascular lesions. They accumulate in subendothelial space through the gaps of endothelial cells of arteriosclerotic and inflamed vessels. The lipid microspheres are much more stable than liposomes and can be mass produced for commercial use. Moreover, there is no toxicity associated with the use of a large amount of lipid microspheres. Prostaglandin E_1, the methyl ester of isocarbacyclin, and some anti-inflammatory drugs were incorporated in lipid microspheres, and it was found that these lipo preparations were significantly superior to each free drug in term of efficacy and safety not only in basic studies but also in clinical trials. Bioactive peptides which could not be incorporated into lipid microspheres were conjugated with a modified lecithin. It was found that the tissue distribution of these lecithinized peptides was somewhat similar to lipo preparations. It is considered therefore that lecithinization of bioactive proteins may be one approach to developing drug delivery systems for these peptides.

When we use a hormone as a drug, we can give it systemically. It enters into the blood stream and acts physiologically and pharmacologically as it is acting in the body. Contrary, autacoids such as prostaglandins and superoxide dismutase are

0097–6156/91/0469–0266$06.00/0
© 1991 American Chemical Society

produced locally in the body and act also locally for their own purposes, and they are rapidly inactivated when they enter systemic circulation to avoid side effects. Therefore, when they are injected systemically, accumulation of these active substances into a diseased site is not satisfactory, and systemic side effects may occur. Therefore, a certain type of drug delivery system is needed for systemic administration of these drugs. Prostaglandin E_1 (PGE_1) and prostacyclin (PGI_2) are known to be very valuable for the treatment of many arteriosclerotic diseases if they can be free from adverse reactions.

An excellent carrier is needed to deliver a sufficient amount of prostaglandins to the diseased site. Liposomes have been studied for a long time as possible drug carriers. However, the clinical use of liposomes has delayed because of some difficulties in mass production, sterilization, stability and safety. Since 1980 we have attempted to use lipid microspheres (lipid emulsions) instead of liposomes as a better carrier for lipophilic drugs (*1*).

In this study we incorporated PGE_1 and a PGI_2 derivative into lipid microspheres, and the tissue distribution and clinical effectiveness of these lipo-preparations were studied.

Bioactive peptides such as superoxide dismutase and interferon are also hoped to be accumulated in the inflamed and vascular lesions. However, these active peptides cannot be incorporated in lipid microspheres. Instead of incorporating them into lipid microspheres, we devised a method to combine the bioactive peptides with a chemically modified lecithin. In this study, we also examined the tissue distribution of lecithinized IgG.

Experimental and Results

Preparation of lipid microspheres. The lipid microspheres (lipo-PGE_1) with a diameter of 0.2 to 0.3 μ m are prepared from the drug, soybean oil and lecithin (Figure 1). The drug to be enclosed in the microspheres is first dissolved in soybean oil, and then emulsified with lecithin by a Manton-Gaulin homogenizer (*1,2*).

Table I shows the composition of lipo-PGE_1. Oleic acid was used to improve the stability of PGE_1. Glycerol was added to make the water phase isotonic. Isocarbacyclin, a prostacyclin derivative (TEI9090), was incorporated into the lipid microspheres (lipo-PGI_2) in a similar composition except for the use of oleic acid.

Tissue distribution of lipo-preparations. The tissue distribution of lipid microspheres in normal and pathologic animals was studied. Research into liposomes of similar size suggested that lipid microspheres accumulated preferentially in the reticuloendothelial system, inflammatory sites, or certain tumors. The distribution of lipid microspheres to these tissues has been found in our studies (*1,2*). Interestingly, our study showed that lipid microspheres accumulated, particularly at high concentrations, in damaged vascular walls such as atherosclerotic vascular walls.

Accumulation of lipid microspheres at the site of vascular lesions can be evaluated by three methods. First, electron microscopy can be used to examine the site of tissue damage. The second method is to follow the delivery of lipid

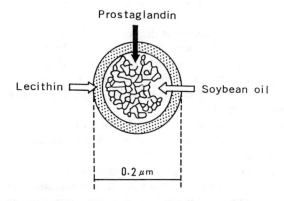

Figure 1. A model of lipo-PGE$_1$.

Table I. Composition of Lipo-PGE$_1$

Prostaglandin E$_2$	5 μg
Soybean oil	100 mg
Egg yolk phospholipids	18 mg
Oleic acid	2.4 mg
Glycerol	22.1 mg
Water for injection	q.s.
Total volume	1 mL

pH: 4.5-6.0
Osmolarity: 280-300 mOsm
Particle size: 200-300 nm in diameter
Shelf life: 12 months at cold room

microspheres by incorporating a radiolabelled compound. The third is scintigraphic clinical assessment of technetium-labelled lipid microspheres. We injected lipid microspheres intravenously into SHR (spontaneously hypertensive rats), and then examined them by electron microscope (*3*). Many lipid microspheres accumulated below the vascular endothelium in the arterial lesions of SHR which had lesions similar to arteriosclerosis. In contrast, lipid microspheres did not accumulate in subendothelial spaces in normal rats or SHR before the development of vascular lesions.

We investigated the delivery of lipid microspheres to atherosclerotic lesions induced by cannulation in rabbits. Many lipid microspheres, which passed through the gaps between the endothelial cells, accumulated below the endothelium at the site of atherosclerotic lesions (*3*). When lipid microspheres incorporating radiolabelled palmitate ester were injected, radioactivity in the atherosclerotic vascular wall was about double the level in normal sites (*3*). Nakura et al. similarly studied the distribution of radioactivity in SHR, and found that lipo-PGE_1 accumulated at the site of pathologic lesions at higher concentrations than free PGE_1 (*4*).

Technetium-labelled lipid microspheres were given to patients with ASO (arteriosclerosis obliterans) by intravenous injection, followed by scintigraphy. Technetium accumulated at the sites corresponding to the atherosclerotic lesions (*5*).

Clinical studies on lipo-PGE_1. After several phase 2 clinical studies were completed (*6,7*), controlled multicentered trials of lipo-PGE_1 were conducted (Table II). In all diseases tested, lipo-PGE_1 was significantly more useful than PGE_1 CD (cyclodextrin) or other reference standard. A cold feeling in the extremities, numbness, paresthesia, and pain were very responsive to lipo-PGE_1. The size of the ulcers which could be assessed objectively was measured in a blind manner. Ulcer lesions associated with collagen diseases regressed significantly within a week of daily treatment with lipo-PGE_1. After 4 weeks of treatment, the difference from the control group was significant (*8*). One double-blind study demonstrated a better safety profile for lipo-PGE_1 than for placebo.

There have been many sporadic reports that lipo-PGE_1 is effective in fulminant hepatitis, neuralgia associated with herpes zoster, multiple spinal canal stenosis, cerebral infarction, myocardial infarction, chronic renal failure, and bed sores as well as for its registered indications.

Lipo-PGE_1 was reported to be 10 to 20 times more effective than PGE_1 in the treatment of ductus dependent congenital heart diseases. It can be used at much lower doses, and accordingly, adverse reactions are reduced (*9*). The drug is approved for this indication, and has already been used safely in many patients.

Lipo-PGI_2. Prostacyclin (PGI_2) is more potent than PGE_1 in antiplatelet and antithrombotic effects, and is expected to be very useful in the treatment of various thrombotic diseases (*10,11*). However, it causes more adverse reactions, such as hypotension and facial flushes, because of its vasodilation and possible suppression of feed-back mechanisms. Many attempts at clinical applications have failed. We have conducted studies on methylated TEI9090 (Figure 2), a chemically stable and

Table II. Summary of Clinical Results on Lipo-PGE$_1$

1. Arterial-duct-dependent congenital heart disease

	No. of Patients	Efficacy	Dose	Adverse Reaction
Lipo-PGE$_1$	83	94.0%	5 ng/kg/min	30.1%
PGE$_1$-CD	(historical)	50~70%	50~100 ng/kg/min	70~80%

2. Buerger's disease plus arteriosclerosis obliterans (double blind trial)

	No. of Patients	Improvement*	Dose	Adverse Reaction**
Lipo-PGE$_1$	62	59.7%	10 μg/day/4 weeks iv	9.4%
IHN	62	40.3%	1200 mg/day/4 weeks po	6.3%

IHN: Inositol Hexa Nicotinate   *$p<0.05$, **N.S.

3. Diabetes-associated peripheral vascular and nervous disorders

	No. of Patients	Improvement*	Dose	Adverse Reaction**
Lipo-PGE$_1$	84	60.7%	10 μg/day/4 weeks iv	12.9%
PGE$_1$-CD	89	38.2%	40 μg/day/4 weeks iv	23.8%

*$p<0.01$, **N.S.

4. Collagen disease-associated peripheral vascular disturbances (double blind trial)

	No. of Patients	Improvement*	Dose	Adverse Reaction**
Lipo-PGE$_1$	66	54.5%	10 μg/day/4 weeks	19.1%
Placebo	65	29.2%		16.4%

*$p<0.001$, **N.S.

5. Vibration disease (double blind trial)

	No. of Patients	Improvement*	Dose	Adverse Reaction**
Lipo-PGE$_1$	49	85.7%	10 μg/day/4 weeks iv	36.8%
PGE$_1$-CD	48	64.6%	40 μg/day/4 weeks iv	58.2%

*$p<0.05$,   **$p<0.5$

lipophilic PGI_2 derivative (*12*), incorporated into lipid microspheres (lipo-PGI_2) as we did with lipo-PGE_1. The methylated TEI9090 could be incorporated into lipid microspheres satisfactorily and was chemically and physically stable (*13*), and so its clinical application seemed to be feasible.

Sim and his co-workers evaluated the antithrombotic effect of lipo-PGI_2 in the hamster's cheek pouch model. Lipo-PGI_2 was 500 times stronger than free TEI9090 in antithrombotic effect (*13*). Therefore, the clinical effect of lipo-PGI_2 seemed to appear at less than 1/100 the doses recommended for PGI_2. Otsu et al. demonstrated that TEI9090 produced more potent vascularization than natural PGI_2 (*14*). As stated above, lipid microspheres are expected to be preferentially delivered to the site of vascularization. These results suggest that lipo-PGI_2 may be very valuable in the treatment of peripheral vascular diseases or cerebral infarction. We made a double-blind comparison of lipo-PGI_2 at a low dose of 2 µg daily with placebo in the treatment of chronic cerebral infarction. The duration of the treatment was short (2 weeks), but lipo-PGI_2 was obviously and significantly more effective (*15*). Both mental and neurological symptoms responded and the patients were better motivated for further therapy.

Lecithinized IgG. As a study on drug delivery system (DDS) for bioactive polypeptides, human immunoglobulin G (IgG) which is chemically stable was used as a model peptide. In order to increase its tissue affinity, IgG was covalently conjugated with an amino-lecithin, which was synthesized by introducing an amino group into the side chain of lecithin at position 2, using water soluble carbodiimide. Sepharose 4B colum chromatographic studies indicated that the IgG combined 46 amino-lecithin residues with the monomeric peptide. The distribution of lecithinized IgG into various tissues and organs of normal mice was investigated. It was shown that the tissue distribution of lecithinized IgG showed similar characteristics to those of lipid microspheres (Figure 3). It produced high concentrations in particular in the kidneys, liver and spleen. When the lecithinized IgG was incubated in vitro with human lymphocytes and cancer cells, it showed a high affinity to the lymphocytes and MM46 and sarcoma 180 cancer cells. Thus, the lecithinization of bioactive peptides is expected to be an effective approach to delivery of the peptides.

Discussion

Lipid microspheres were used instead of liposomes as a carrier of lipophilic drugs in this study. Because the outside layer of liposomes and lipid microspheres is lecithin the distribution into the body was expected to be similar.

A large amount of lipid microspheres has been used under the name of Intralipid for almost 50 years and we used only 1 mL of it for DDS. Therefore, there was no problem with regard to safety, stability, sterilization and mass production.

It was found in animal and clinical studies that lipid microspheres accumulated particularly in arteriosclerotic and damaged vessel walls. Earlier studies (*16,17*) of lipid emulsions demonstrated also that lipid microspheres had an affinity to vascular walls, including capillaries, like chylomicrons. Shaw et al. reported that they had

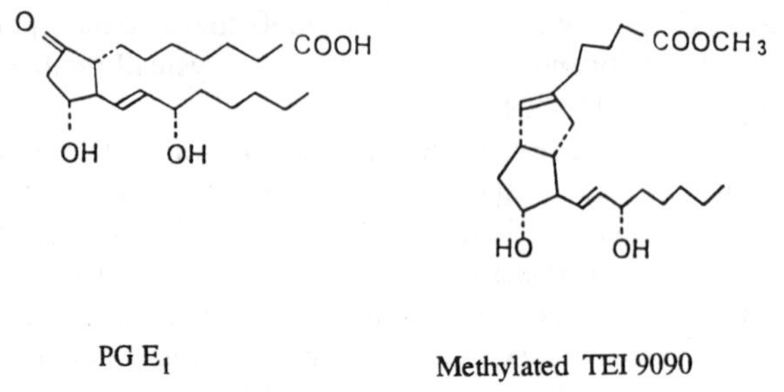

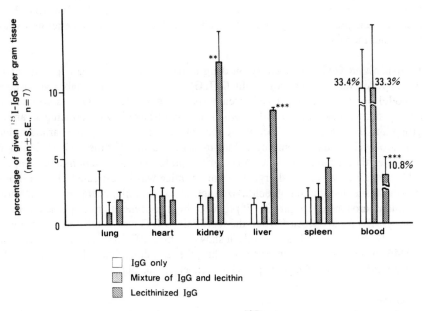

PG E₁   Methylated TEI 9090

Figure 2. Chemical structure of prostanoids used for LM-preparation.

Figure 3. Tissue distribution of lecithinized ^{125}I-IgG (specific activity: 0.01 μ Ci/μ g) in C_3H mice (4 hours after 200 μ L/mouse iv injection) (**p<0.01 ***p<0.001 in relation to IgG only and IgG-lecithin mixture).

more affinity to vascular walls at inflamed sites (18). From our studies using an electromicroscope, it was shown that lipid microspheres with a diameter of 0.2 μm passed through the gaps between endothelial cells or through endothelial cells, and accumulated below them.

As described above, lipo-PGE₁ is delivered preferentially to the site of vascular lesion, and PGE_1 becomes less irritating, because it is incorporated into the microspheres. In addition, inactivation of PGE_1 in the lungs is reduced for the same reason. All these features of the preparation strongly suggest that lipo-PGE_1 would be very valuable not only in animal models but also in humans. Sim et al. found in a hamster's cheek pouch test (19,20) that lipo-PGE_1 was better for thrombolysis than free PGE_1 (*21*). In an animal model of lauric acid-induced peripheral vascular disease, better effectiveness was obtained with lipo-PGE_1 than with free PGE_1 (*22*).

Yokoyama et al. compared the hypotensive effect of lipo-PGE_1 with that of free PGE_1 CD in diabetic and spontaneously hypertensive rats (SHR). The reactivity to vasoactive substances decreased as the diabetes progressed. This phenomenon was found in both PGE_1 CD and isoproterenol, while response to lipo-PGE_1 was enhanced as the diabetes progressed (*23,24*). The most striking difference between lipo-PGE_1 and PGE_1 CD was found in diabetic rats aged 10 weeks. In these animals, the hypotensive effect of lipo-PGE_1 was about 25 times the level of PGE_1 CD. In the SHR model, lipo-PGE_1 was shown to be much more potent than PGE_1 CD.

Of the various pharmacological actions of PGE_1 and PGI_2, one action which has not yet been studied intensively, but seems to be important for the treatment of arteriosclerotic diseases, is their effect on vascularization. This suggests the need to determine if lipid microspheres accumulate at the site of vascularization. We are now conducting a study to investigate this possibility. Since the space between endothelial cells in new blood vessels is as big as that of vessels with sclerosis or inflammation, lipid microspheres are very likely to accumulate in new blood vessels. Figure 4 shows a schema of the distribution and accumulation of lipid microspheres at the site of vascular lesions.

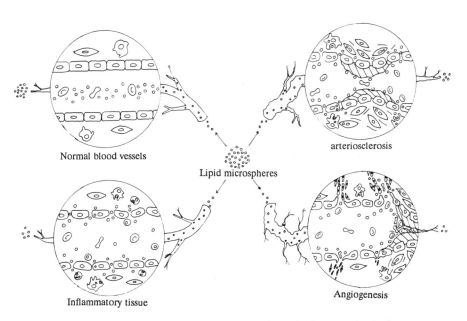

Figure 4. Distribution and accumulation of LM in the vascular lesions.

Acknowledgments

The authors wish to thank the Green Cross Corporation, Taisho Pharmaceutical Co., Ltd., Drs. Hoshi and Yanagawa of St. Marianna University and Mrs. Sugi for their cooperation in this study and preparing the manuscript.

Literature Cited

1.   Mizushima Y, Hamano T, Yokoyama K. *Ann. Rheum. Dis.* **1982**,*41*, 263.
2.   Yokoyama K, Ueda Y, Kikukawa A, et al. In *Targeting of Drugs: Optimization Strategies.* G. Gregoriadis, Ed., Plenum: NY 1990 (in press).
3.   Mizushima Y, Hamano T, Haramoto S, et al. *Prostagl. Leukotr. Essent. Fatty Acids* **1990**, *41*, 269.
4.   Nakura K, Hamano T, Shintome M, et al. *Basic Clin. Res. (in Japanese)* **1986**, *20*, 143.
5.   Kiyokawa S, Igarashi R, Iwayama T, et al. *Jap. J. Inflamm. (in Japanese)* **1987**, *7*, 551.
6.   Mizushima Y, Yanagawa A, Hoshi K. *J. Pharm. Pharmacol.* **1983**, *35*, 666.
7.   Hoshi K, Mizushima Y, Kiyokawa S, et al. *Drugs Exp. Clin. Res.* **1986**, *12*, 681.
8.   Mizushima Y, Shiokawa Y, Homma M, et al. *J. Rheum.* **1987**, *14*, 97.
9.   Momma K. *Int Angiol.* **1984**, *3 Suppl.*, 33.
10. Gryglewski R.J. *Adv. Prostagl. Trombox. Leukotr. Res.* **1985**, *15*, 539.
11. Grose R, Greenberg M, Strain J, et al. *Am. J. Cardiol.* **1985**, *55*, 1625.
12. Shibasaki M, Torisawa Y, Ikegami S, et al. *Tetrahedron Lett.* **1983**, *24*, 3493.
13. Mizushima Y, Igarashi R, Hoshi K, et al. *Prostagl.* **1987**, *33*, 161.
14. Ohtsu A, Fujii K, Kurozumi S. *Prostagl. Leukotr. Essent. Fatty Acids* **1988**, *33* 35.
15. Hoshi K, Mizushima Y. *Prostagl.* **1990**, *40* 155.
16. Davis S.S., Hadgraft J, Palin K.J. In *Encyclopedia of Emulsion Technology 3*, Becher P, Ed., Dekker, New York, 1985.
17. Hallberg D. *Acta. Physiol. Scan.* **1965**, *65 Suppl*, 254.
18. Shaw I.H., Knight C.G., Thomas D.P.P., et al. *Brit. J. Exp. Pathol.* **1979**, *60*, 142.
19. Duling B.R., Berne R.M., Born G.V.R. *Microvascul. Res.* **1968**, *1*, 158.
20. Sim A.K., Uzan A. *Arzneimittel-Forschung/Drugs Res.* **1979**, *29* 508.
21. Sim A.K., McCraw A.P., Cleland M.E., et al. *Arzneimittel-Forschung/Drugs Res.* **1986**, *36* 1206.
22. Otomo S, Mizushima Y, Aihara H, et al. *Drugs Exp. Clin. Res.* **1985**, *11*, 627.
23. Yokoyama K, Okamoto H, Watanabe M, et al. *Drugs Exp. Clin. Res.* **1985**, *11*, 611.
24. Hamano T, Shintome M, Watanabe M. *Basic Clin. R. (in Japanese)* **1986**, *20*, 93.

RECEIVED March 19, 1991

Chapter 24

Drug Delivery by Organ-Specific Immunoliposomes

Kazuo Maruyama[1], Atsuhide Mori[1], Stephen J. Kennel[2],
Marjan V. B. Waalkes[3], Gerrit L. Scherphof[3], and Leaf Huang[1,4]

[1]Department of Biochemistry, University of Tennessee, Knoxville, TN 37996
[2]Biology Division, Oak Ridge National Laboratory, P.O. Box Y,
Oak Ridge, TN 37830
[3]Department of Physiological Chemistry, University of Groningen,
Groningen, Netherlands

Monoclonal antibodies highly specific to the mouse pulmonary endothelial cells were conjugated to liposomes. The resulting immunoliposomes showed high levels of lung accumulation when injected intravenously into mice. Optimal target binding and retention were achieved if the lipid composition included ganglioside (GM_1) to reduce the uptake of immunoliposomes by the reticuloendothelial system. Details of the construction and optimization of these organ-specific immunoliposomes are reviewed. The drug delivery potential of this novel liposome system was demonstrated in an experimental pulmonary metastasis model. Immunoliposomes containing a lipophilic prodrug of deoxyfluorouridine effectively prolonged the survival time of the tumor-bearing mice. This and other therapeutic applications of the immunoliposomes are discussed.

Liposomes are versatile and efficient carriers for both hydrophilic and hydrophobic drugs. They are biocompatible, biodegradable and usually non-immunogenic. Drugs encapsulated in liposomes are protected from the action of the external enzymes and other inactivation factors. However, liposomes injected systemically localize predominantly in the mononuclear phagocytes of the reticuloendothelial system (RES) primarily in the liver and secondarily in the spleen (1). This natural homing activity to the RES, or so-called "passive targeting", has been exploited to deliver drugs to these cells. For example, antiparasitics encapsulated in liposomes effectively kill the microorganisms residing in the macrophages (2,3). Liposomes containing the macrophage-activating factors are also useful for both antitumor and antiviral activities (4).

[4]Corresponding author.

0097–6156/91/0469–0275$06.00/0
© 1991 American Chemical Society

Delivery of liposomes to cells, tissues or organs other than the RES is referred to as the "active targeting". It is necessary to conjugate a suitable targeting ligand, such as a monoclonal antibody, to the liposome surface for the target specificity. There are many established methods of preparing the antibody-coated liposomes, i.e. immunoliposomes, and the target-binding specificity of the immunoliposomes has been demonstrated in many cell-free, or tissue culture systems (5). However, efficient target binding and retention of the immunoliposomes in an animal model, particularly those bearing tumors at a specific site, have not been demonstrated until recently (6,7,8). Two different problems have been identified: RES uptake of the immunoliposomes and failure of immunoliposomes to cross the anatomical barriers (9). The former problem has been solved by the recent development of liposomes with a reduced RES affinity. These liposomes can escape the surveillance of the macrophage (10-13). In this report, we will describe the importance of the prolonged liposome circulation time for optimal target binding and retention of immunoliposomes (14). The latter problem stems from the fact that liposomes or immunoliposomes are too large to penetrate the endothelial barrier of the vessel wall. Unfortunately many tumor cells, either in a primary or metastatic tumor, are not readily accessible by liposomes. Thus, a direct targeting of immunoliposomes to the tumor cells is probably difficult. As an alternative approach, we have decided to design an organ-specific drug delivery system which allows an efficient binding and retention of immunoliposomes to the vascular endothelial cells of that organ. Drugs released from the bound immunoliposomes need only diffuse a short distance to reach the tumor cells in that organ. This concept is described schematically in Figure 1.

The mouse pulmonary metastatic tumor lends itself as an ideal model to test the new concept. Experimental metastasis can be easily established by iv injection of free tumor cells which usually form numerous metastatic tumor nodules in the lung. It has been reported that these tumors are either poorly vascularized or not at all (15). One of us (S.J.K.) has isolated two rat monoclonal IgG antibodies which specifically bind with a glycoprotein antigen, gp112, located on the lumenal surface of the capillary endothelial cells in the mouse lung (16). The gp112 is highly organ-specific; its concentration in the lung is at least 34-fold greater than at any other organs in the mouse (15). We have used this model system to test the drug delivery potential of the organ-specific immunoliposomes.

Materials and Methods

Materials. Egg phosphatidylcholine (PC), bovine brain phosphatidylserine (PS) were obtained from Avanti Polar Lipids Inc. (Birmingham, AL) and cholesterol was from Sigma (St. Louis, MO). Ganglioside GM_1, bovine, was obtained from Calbiochem (San Diego, CA). Diethylenetriamine pentaacetic acid distearylamide complex (DPTA-SA) was synthesized according to ref. 17 and ^{111}In-DTPA-SA was prepared as described (7). This lipophilic radiolabel is not transferred to the serum components from liposomes (unpublished data), nor is it rapidly metabolized in vivo (7). The synthesis of N-(glutaryl)phosphatidylethanolamine(NGPE) has been described (18). Dipalmitoyl deoxyfluorouridine(dpFUdR) was synthesized as described (24).

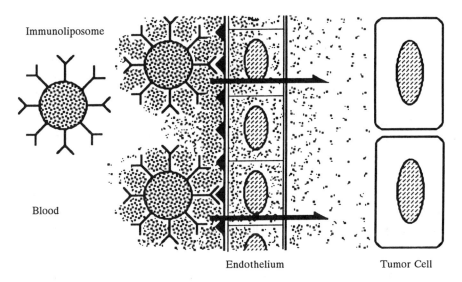

Figure 1. Schematic description of organ-specific immunoliposomes.

Antibody. Rat monoclonal antibody 34A was purified from nu/nu mouse ascites fluid as described (*19*). The 34A was radiolabeled with ^{125}I using IDO-GEN (Pierce, Rockford, IL) method to a specific activity of 2 to 4 x 10^5 cpm/μ g, and conjugated with NGPE as previously described (*7*).

Preparation of immunoliposomes. The 34A-immunoliposomes were prepared by a detergent dialysis method as previously described (*7*). Lipid mixtures (with or without dpFUdR) containing ^{111}In-DTPA-SA (1%) were dried with N_2 gas. The dried lipid was solubilized with 100 mM octylglucoside in PBS (lipid/octylglucoside = 1:5, mole/mole). The resultant solution was mixed vigorously with 34A conjugated with NGPE, and then the mixture was dialyzed against PBS for 16-20 h at 4°C. The immunoliposomes were extruded 4 times through a stack of two Nuclepore membranes (pore size 0.4 μ m). The average size of liposomes was measured using a Coulter N4SD sub-micron particle size analyzer (Hialeah, FL). The immunoliposomes were separated from the unbound antibody (50-70% of the original amount) on a Bio-Gel A1.5M (Bio-Rad) column. The peak liposome fractions were pooled and diluted to 1 mg lipid/mL PBS (pH 7.4). The amount of 34A incorporation into liposomes and the final ratio of antibody to lipid in the 34A-immunoliposomes were calculated from the radioactivities of ^{111}In for lipids and ^{125}I for the antibody.

Biodistribution studies. Liposomes (200 μ g lipid) were injected into Balb/c mice (6-8 weeks old) via the tail vein and the distribution of the liposomes in major organs and blood were examined as a function of time. At the desired time, mice were anesthetized lightly and bled by eye puncture. Blood was collected and weighed. The mice were sacrificed by cervical dislocation and dissected. Organs were collected, weighed, and analyzed for ^{111}In radioactivity in a gamma counter. The results were presented as percent of the total injected dose for each organ. The total radioactivity in the blood was determined by assuming that the total volume of blood was 7.3% of the body weight (*20*).

Tumor cells. EMT6 cells were grown as a monolayer culture in DMEM medium containing 20% fetal calf serum (*21*). Cells were detached from the plate by trypsin-EDTA treatment and washed in PBS. A total of 5 x 10^3 cells were injected per mouse via the tail vein of Balb/c mice (6-8 weeks old) to induce experimental lung metastatic tumors. Immunoliposomes were injected iv 2 and 4 days after the tumor cell injection. The survival of mice was followed over the next 60 days.

Results and Discussion

Optimization of Immunoliposome Binding to the Lung. In order to reduce the uptake of immunoliposomes by the liver and spleen, we have tested the effect of incorporating GM_1 into the liposome membrane on the lung binding of immunoliposomes. We (*14,22*) and others (*10,11,12*) have demonstrated that the presence of this monosialoganglioside in the liposomes effectively reduces the RES uptake and prolongs the circulation time of liposomes. Since GM_1 contains one negative charge per molecule, we have included the negatively charged phosphatidylserine (PS) in the control liposomes. Also shown in Figure 2 is the effect of the antibody 34A to lipid ratio of immunoliposomes on the target binding. The 34A-liposomes of a high antibody density bound rapidly to the lung; it reached the steady-state level within 5 min of injection. The 34A-liposomes of lower antibody densities bound less rapidly and showed a lower steady-state level. Liposomes containing no 34A did not bind to the lung at all. It is also clear from the data in Figure 2 that 34A-liposomes containing GM_1, bound more effectively to the lung target than the ones containing PS. In a separate experiment, it was shown that the liver uptake rate of the 34A-liposomes containing GM_1 was approximately 10-fold lower than the ones containing PS (*14*). Thus, immunoliposomes with a reduced RES affinity bind to the target more effectively than the ordinary immunoliposomes.

These data indicate two kinetically competing processes: binding to the lung target and uptake by the RES. The steady-state level of the lung accumulation of immunoliposomes is determined by the relative rates of the two processes. Increasing the antibody density of immunoliposomes favors an enhanced rate of target binding. Inclusion of GM_1 in the liposome membrane effectively reduces the rate of RES uptake of immunoliposomes. When the two factors are optimized at the same time, we can routinely obtain 70-75% of the injected dose accumulating in the lung within 15 min of injection.

Target Retention of Immunoliposomes. Successful therapy mediated by the organ-specific immunliposomes requires a relatively long residence time of the immunoliposome in the target organ. Figure 3 shows that 34A-liposomes containing GM_1 stayed in the lung longer than the ones containing PS. Since PS-containing liposomes show a high affinity to macrophages, this result suggests the removal of immunoliposomes from the lung may involve the action of macrophages. Alternatively, released immunoliposomes could reenter the circulation and rebind to the target if they are not taken up by the RES. GM_1-containing 34A-liposomes would show a greater chance of rebinding than the ones containing PS due to the difference in the RES affinity. Also shown in Figure 3 is that 34A-liposomes more enriched with antibody were retained longer by the target than the ones less enriched. This is understandable because immunoliposomes containing higher numbers of antibody are more multivalent than the ones containing less antibody. Multivalent binders dissociate very slowly from the receptors (*23*). These data have taught us that for the tumor therapy, we should use immunoliposomes of high antibody density which also contain GM_1 in the liposome membrane. This endows the liposomes with a relatively high resistance to the RES uptake.

An Experimental Pulmonary Metastasis Model. EMT6 cells were originally derived from a spontaneous breast tumor of a Balb/c female mouse (*21*). These cells grow rapidly in tissue culture and can be transplanted subcutaneously to form solid tumors in mice. When 5×10^3 single cells were injected via the tail vein, the mice developed massive tumor nodules in the lung and usually died after 20-22 days. No growth of the tumor was seen in any other organs. The tumor-bearing mice showed a

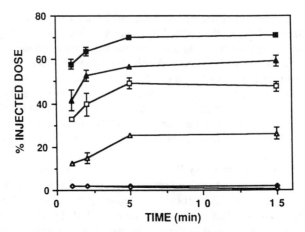

Figure 2. Lung uptake of GM_1-containing or PS-containing liposomes. Liposomes (200μg lipid) were injected via tail vein. Bar is S.D. (n=3). Data taken with permission from reference 14. Key: ■, 34A-liposomes (PC:chol:GM_1=10:5:1), Ab:lipid=1:11 (w/w), 297 nm in average diameter; ▲, 34A-liposomes (PC:chol:GM_1=10:5:1), Ab:lipid=1:37 (w/w), 292 nm in average diameter; ◆, liposomes (PC:chol:GM_1=10:5:1), without antibody, 268 nm in average diameter, □; 34A-liposomes (PC:chol:PS=10:5:1), Ab:lipid=1:8 (w/w), 253 nm in average diameter; △, 34A-liposomes (PC:chol:PS=10:5:1), Ab:lipid=1:31 (w/w), 255 nm in average diameter; ◇, liposomes (PC:chol:PS=10:5:1), without antibody, 235 nm in average diameter.

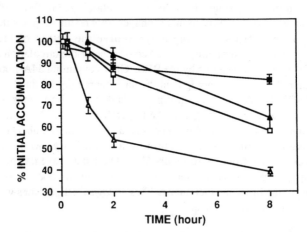

Figure 3. Retention of immunoliposomes in lung. Immunoliposomes (200 μg lipid) labeled with ^{111}In-DTPA-SA were injected iv. The percent initial accumulation in lung was calculated at indicated time intervals. Bar is S.D. (n=3) Data taken with permission from reference 14. Key: ■, 34A-liposomes (PC:chol:GM_1=10:5:1), Ab:lipid=1:11 (w/w), 297 nm in average diameter; ▲, 34A-liposomes (PC:chol:GM_1=10:5:1), Ab:lipid=1:37 (w/w), 292 nm in average diameter; □, 34A-liposomes (PC:chol:PS=10:5:1), Ab:lipid=1:8 (w/w), 253 nm in average diameter; △, 34A-liposomes (PC:chol:PS=10:5:1), Ab:lipid=1:31 (w/w), 255 nm in average diameter.

high level of 34A-liposome accumulation in the lung, indistinguishable from the normal mice (data not shown). We have used this experimental model to test the ability of 34A-liposomes to deliver cytotoxic drugs for tumor therapy.

Choice of the Drug. Organ-specific immunoliposomes do not directly deliver the encapsulated drug to the tumor cells. Instead, the drug has to diffuse from the bound liposomes to the nearby tumor cells. We have chosen a hydrophobic drug, i.e. 3',5'-0-dipalmitoyl-2'-deoxyfluorouridine (dpFUdR), because a hydrophobic drug is likely to diffuse more efficiently across the cellular barriers. Furthermore, nearly all of the added dpFUdR was incorporated into the 34A-liposome membranes. Its presence in the liposome membrane did not interfere with the incorporation of the 34A antibody into the same liposomes. The dpFUdR is rapidly metabolized to FUdR which is cytostatic by inhibiting the cellular DNA synthesis (Figure 4). Therefore dpFUdR is a lipophilic prodrug of FUdR (*24*). EMT6 cells in culture could be killed by liposomal dpFUdR with a LD_{50} of 7 ng/mL, indicating that it is a potent cytostatic drug for this tumor system. Waalkes and Scherphof have shown that dpFUdR formulated in antibody-free liposome is extraordinarily toxic to mice, due to the high concentrations of the drug delivered to the liver and the slow elimination of the drug by the animal (*25*). This drug is, however, suitable for the 34A-liposome system, because the immunoliposomes are supposed to direct the drug away from the major organ which converts the prodrug into the drug, i.e. the liver.

Tumor Therapy by Organ-specific Immunoliposomes. Immunoliposomes containing PC:chol:GM_1:dpFUdR = 10:5:1:0.5 and 34A antibody (protein/lipid ratio = 1:17, w/w) were prepared by a detergent-dialysis method (see materials and methods). Immunoliposomes containing dpFUdR were injected iv at 2 mg drug/kg body weight 2 and 4 days after tumor cell injection. The survival time of the treated mice was measured over a 60-day period. Figure 5 shows the therapeutic effect of the immunoliposomes. It is clear that mice receiving the 34A-liposomes with this drug showed a significantly prolonged mean survival time (MST) of 33 days. Mice treated with free drug formulated in a vegetable oil emulsion, or with the drug-loaded liposomes without the antibody, or with 34A-liposomes containing drug but no GM_1, did not show any therapeutic effects as their MST's were indistinguishable from those treated with PBS, i.e. approximately 20-22 days. These data indicate that the organ-specific immunoliposomes are effective for the target-specific therapy of a metastatic tumor. These results, although preliminary, are very encouraging.

Other Potential Advantages of Organ-specific Immunoliposomes. Because the immunoliposomes are designed to target to the normal, healthy cells, and are not tumor cell-specific, this drug delivery system should be theoretically effective in treating any tumor cells grown in the lung, thus providing a general delivery system for the organ. The drug delivery activity of the immunoliposome should not be limited to the anticancer drugs; drugs efficacious in treating other pulmonary diseases can also be delivered. In fact, the same 34A-liposome system effectively delivers acyclovir, an antiviral drug, for the treatment of an experimental viral pneumonia

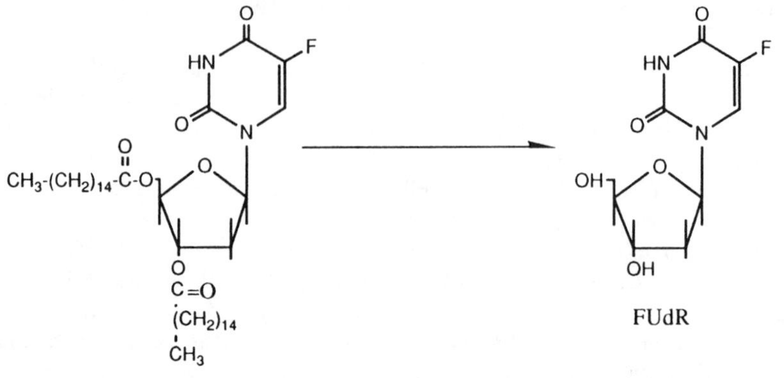

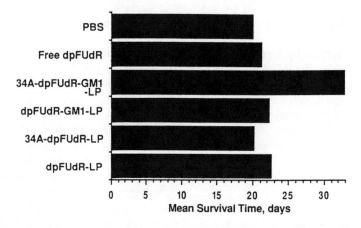

3',5'-O-dipalm-FUdR

Figure 4. Structure and metabolism of dpFUdR.

Figure 5 Mean survival time of mice treated with dpFUdR in different liposome formulations. PBS=phosphate buffered saline, free dpFUdR=free drug formulated in a vegetable oil-tween emulsion, 34A-dpFUdR-GM_1-LP=34A-liposomes containing PC:chol:GM_1:dpFUdR (10:5:1:0.5), dpFUdR-GM_1-LP=liposomes containing PC:chol:GM_1:dpFUdR (10:5:1:0.5), 34A-dpFUdR-LP=34A-liposomes containing PC:chol:dpFUdR (10:5:0.5), dpFUdR-LP=liposomes containing PC:chol:dpFUdR (10:5:0.5). The protein-to-lipid ratios of both 34A-containing liposome preparations were approximately 1:17, w/w. Five to 6 animals per group were used. Bar is standard deviation. Value for 34A-dpFUdR-GM_1-LP is significantly different from all others, $p < 0.01$.

model (unpublished data). These advantages will be explored further in the future experiments.

List of Abbreviations

Chol:	cholesterol
dpFUdR:	3',5'-0-dipalmitoyl-2'deoxy-fluorouridine
GM_1:	ganglioside GM_1
MST:	mean survival time
NGPE:	N-glutaryl phosphatidylethanolamine
PBS:	phosphate buffered saline (137 mM NaCl, 2.7 nM KCl, 1.5 mM KH_2PO_4, 0.1 mM Na_2HPO_4, pH 7.4).
PC:	egg phosphatidylcholine
PS:	phosphatidylserine
RES:	reticuloendothelial system
[111]IN-DTPA-SA:	[111]Indium diethylenetriamine pentaacetic acid distearylamide complex

Acknowledgment

We thank Carolyn Drake for preparing the manuscript. The work was supported by NIH grant CA24553 to L.H. S.J.K. was supported under contract DE-AC05-84OR21400 with the Martin Marietta Energy Systems, Inc.

Literature Cited

1. Hwang, K.J. In *Liposomes;* Ostro, M.J., Ed., Marcel Dekker Inc., New York, NY, 1987, pp. 109-156.
2. Alving, C.R.; Steck, E.A.; Chapman, W.L.; Waits, V.B.; Hendricks, L.D.; Swartz, G.M.; Hanson, W.L. *Proc. Natl. Acad. Sci. USA*, **1978**, *75*, 2959.
3. New, R.R.C.; Chance, M.L.; Heath, S. *J. Antimicrobial Chemother.*, **1981**, *8*, 371.
4. Koff, W.C.; Showalter, S.D.; Hounpar, D. Fidler, I.J. *Science*, **1985**, *228*, 495.

5. Wright, S.; Huang, L. *Advanced Drug Delivery Reviews* , **1989**, *3*, 343.
6. Hughes, B.J.; Kennel, S.; Lee, R.; Huang, L. *Cancer Res.*, **1989**, *49*, 6214.
7. Holmberg, E.; Maruyama, K.; Litzinger, D.C.; Wright, S., Davis, M.; Kabalka, G.W.; Kennel, S.J.; Huang, L. *Biochem. Biophys. Res. Commun.*, **1989**, *165*, 1272.
8. Maruyama, K.; Holmberg, E.; Kennel, S.J.; Klibanov, A.; Torchilin, V.P.; Huang, L. *J. Pharm. Sci.*, **1990**, *79*, 978.
9. Poste, G.; Kirsh, R.; *Bio/Technology* , **1983**, *1*, 869.
10. Allen, T.M.; Chonn, A.; *FEBS Lett.*, **1987**, *223*, 42.
11. Gabizon, A.; Papahadjopoulos, D.; *Proc. Natl. Acad. Sci. USA*, **1988**, *85*, 6949.
12. Allen, T.M.; Hanson, C.; Rutledge, J. *Biochim. Biophys. Acta*, **1989**, *981*, 27.
13. Allen, T.M. In *Liposomes in the Therapy of Infectious Diseases and Cancer*, Berestein, G.L.; Fidler, I.J., Eds.; Alan R. Liss, Inc., New York, NY, 1989, pp. 405-415.
14. Maruyama, K.; Kennel, S.J.; Huang, L. *Proc. Natl. Acad. Sci. USA*, **1990**, *87*, 5744.
15. Kennel, S.J., Lee, R., Bultman, S. and Kabalka, G., **1990**. Nucl. Med. Biol. 17, 193-200.
16. Kennel, S.J.; Lankford, T.; Hughes, B.; Hotchkiss, J. *Lab. Invest.*, **1988**, *59*, 692.
17. Kabalka, G.W.; Buonocore, E.; Hubner, K.; Moss, T.; Norley, N.; Huang, L. *Radiology*, **1987**, *163*, 255.
18 Weissig, V.; Lasch, J.; Klibanov, A.L.; Torchilin, V.P. *FEBS Lett.*, **1986**, *202*, 86.
19. Kennel, S.J.; Foote, L.J.; Lankford, P.K. *Cancer Res.*, **1981**, *41*, 707.
20. Wu, M.S.; Robbins, J.C.; Bugianesi, R.L.; Ponpipom, M.M.; Shen, T.Y. *Biochem. Biophys. ACTA* , **1977**, *674*, 255.
21. Rockwell, S. *Lab. Animal Sci.*, **1977**, *27*, 831.
22. Liu, D.X.; Huang, L. *Biochem. Biophys. Acta*, **1990**, *1022*, 348.
23. Bell, G.I. *Science*, **1978**, *200*, 618.
24. Supersaxo, A.; Rubas, W.; Hartmann, H.R.; Schott, H.; Hengartner, H.; Schwendener, R.A. *J. Microencapsulation* , **1988**, *5*, 1.
25. Waalkes, M.V.B.; Scherphof, G.L. *Selective Cancer Therapeutics*, **1990**, *6*, 15.

RECEIVED March 19, 1991

Chapter 25

Novel Polymerized Vesicles with Hydrolyzable Linkages

William J. Bailey[1] and Lin-Lin Zhou[2]

Department of Chemistry and Biochemistry, University of Maryland, College Park, MD 20742

A novel polymerized vesicular system for controlled release, which contains a cyclic α-alkoxyacrylate as the polymerizable group on the amphiphilic structure, has been developed. These lipids can be easily polymerized through a free radical process. It has been shown that polymerization improves the stabilities of the synthetic vesicles. In the aqueous system the cyclic acrylate group, which connects the polymerized chain and the amphiphilic structure, can be slowly hydrolyzed to separate the polymer chain and the vesicular system and generate a water-soluble biodegradable polymer. Furthermore, in order to retain the fluidity and to prepare the polymerized vesicles directly from prepolymerized lipids, a hydrophilic spacer has been introduced.

Phospholipids or similar water-insoluble amphiphilic natural substances aggregate in water to form bilayer liquid crystals which rearrange when exposed to ultrasonic waves to give spherical vesicles. Natural product vesicles are also called liposomes. Liposomes, as well as synthetic bilayer vesicles, can entrap substances in the inner aqueous phase, retain them for extended periods, and release them by physical process.

The most important application recently developed for synthetic liposomes is as potential drug carriers for controlled release, especially for cancer chemotherapy (1). In general, the success of liposomes as vehicles for the transport of specific drugs will largely depend on their stability under physiological conditions. Unlike the naturally occurring membranes, the synthetic vesicles have very limited stability, and this is a

---

[1]Deceased.

[2]Corresponding author: Institute for Environmental Chemistry, National Research Council, Ottawa K1A OR6, Canada

0097–6156/91/0469–0285$06.00/0

serious drawback of these bilayer vesicles for real applications, long-term use, and kinetic studies of permeability.

In the 1980s, polymerization was introduced to overcome the limited stability of synthetic vesicles (2-4). It was found that the stability of the polymerized vesicles was improved dramatically compared to the unpolymerized vesicle and that entrapped substances are released to a much smaller extent from polymerized liposomes than from monomeric ones.

Although polymerization brings many desirable properties into the synthetic liposome system, problems may emerge with the polymer chains on the lipids in certain applications, especially in the use as a drug carrier for controlled release. It is necessary that the polymer chains on the vesicles be cleaved or removed at some point in a biological system in order to release the entrapped substances. So far most of the polymer chains which have been introduced contain carbon-carbon bonds on the backbone of polymers. These carbon-carbon bonds are very inert in biological system and are very difficult to cleave by biological and chemical methods. Also high molecular weight nonbiodegradable polymers remaining inside the body can interfere with biological functions. These problems will definitely hinder the real applications of polymerized vesicles as drug carriers.

Several efforts have dealt with these specific problems. Regen, et al., reported the synthesis and characterization of phospholipids bearing thiol groups (5,6). These phospholipids were used to construct vesicles that could be oxidatively polymerized. The resulting disulfide in the backbone of the polymer could also be reductively depolymerized. Since the disulfide moiety is common to many biopolymers, they suggested that the disulfide-based polymerized vesicles may be biodegradable. Another important approach was introduced by Ringsdorf, et al (7). They synthesized amphiphilic lipids which contained amino acid groups. The amino acid groups on vesicles could be polymerized through peptide bonds. They also suggested that these peptide liposomes should have the advantage of being biodegradable.

However, these approaches are still far from satisfactory. There are no reports about polymerized vesicles which are biodegradable, even though some potentially biodegradable bonds have been introduced into the vesicle system. It should be realized that the introduction of a simple biodegradable linkage into a complex system does not mean that inside the complex system this linkage remains biodegradable. Therefore, novel approaches have to be developed to obtain target-guided drug carriers.

In our laboratories, a series of new cyclic α-alkoxyacrylates has been synthesized and studied (8). These monomers are very reactive in free radical polymerizations and could be easily polymerized to give polymers containing cyclic acetal groups. Since these cyclic acetal groups can be hydrolyzed to give water-soluble poly(α-hydroxyacrylic acid), which has been found to be biodegradable (9,10), it has been reasoned that introduction of this monomer as the monomeric group and utilization of the resulting hydrolyzable acetal as a connecting group between the amphiphilic structure and the polymer chain would give the vesicle reasonable stability and make it possible to remove the polymer chain from the vesicular system. For these reasons, two polymeric lipids containing the cyclic acrylate groups as the polymerizable groups have been synthesized. The principle is shown in Scheme I.

Experimental

Preparation of Vesicles. The tube used for the formation of the vesicular system was first washed with detergent and then rinsed consecutively with distilled water, highly purified water, highly purified methanol and chloroform, and finally dried in an oven overnight. In the tube 2 mg of the lipid was dissolved in 1 mL of chloroform. The solvent was slowly removed by evaporation at reduced pressure leaving the lipid coated on the wall of the tube. The tube was evacuated at 0.1 mm Hg overnight. After 2 mL of highly purified water was added, the suspension was placed in a heating bath with temperature set at 60°C for 3 minutes, and then the suspension was vortex mixed for about 1 minute; this procedure was repeated three times after which the system became a milky suspension. This milky suspension was further dispersed in a bath-type sonication to give an optically clear vesicle system.

Polymerization of Lipid 1 by UV Irradiation. The vesicle suspension prepared as described above was transfered into a quartz tube which was then flushed with nitrogen gas for about 20 minutes. After the tube was sealed with a rubber stopper, it was put on a rotator contained in a miniphotoreactor for UV irradiation for 3 minutes with slow rotation.

Polymerization of Lipid 2 by Free Radical Initiation. Into a polymerization tube was placed 110 mg of Lipid 2, 2 mg of AIBN and 1 mL of benzene. After the tube was evacuated and flushed three times with nitrogen gas and sealed under vacuo, the tube was heated overnight at 60°C. After the sealed tube was opened, the mixture was dissolved in a small amount of chloroform, and this chloroform solution was added to a large excess of methanol. The precipitated polymer was collected by filtration. After the solid was dried in vacuo, 78 mg (71%) of polymer was isolated.

Results and Discussion

The first desired polymeric lipid synthesized in our laboratories contains two long alkyl chains attached to a quarternary nitrogen atom which serves as a polar head group. The polymeric group is placed at the end of one of the alkyl chains through a ester linkage. The major advantage of using a quarternary nitrogen atom as the polar head group is its simplicity in the chemical synthesis. The synthetic route for the preparation of the lipid is outlined in Scheme II. Details of the synthesis have been published elsewhere (*11*). The resulting monomeric lipid was found to be quite stable when kept at 0°C. Because of its amphiphilic structure it can dissolve in many common organic solvents.

Usually the polymerization of the vesicle can be performed by either UV irradiation or by thermal free radical initiators, such as AIBN. However, in the present case, the thermal polymerization of the vesicles was not successful because the hydrolysis of the monomeric acetal group proceeded very quickly at high temperatures. The hydrolysis during polymerization was monitored by the UV spectra of the polymerizing system as shown in Figure 1. The absorption around 240 nm, which is attributed to the cyclic ketene acetal group and neighboring phenyl ester group, shifted to higher wavelength with increasing time; this shift was due to the hydrolysis of the cyclic acetal groups to the benzaldehyde ester functional group. Therefore, polymerization via UV irradiation was the only convenient route to these polymeric vesicles. After 3 minutes of UV irradiation in a miniphotoreactor the sample had no visible change and remained optically clear. The electron microscope picture showed that the shape and size of the vesicles were the same as those before irradiation.

The polymerization through the cyclic acrylate carbon-carbon double bond was confirmed by both the UV spectra of the vesicle suspension and the IR spectrum of the polymerized lipids obtained by removal of the water through evaporation at low temperatures. As can been seen in Figure 2, the UV absorption peak at 238 nm, which is attributed to the cyclic ketene acetal group and the neighboring phenyl ester group, decreased about 35% after 2 minutes of irradiation of the vesicles as the result of the polymerization of the carbon-carbon double bonds of the cyclic ketene acetal. Further irradiation did not decrease the absorption significantly. In the IR spectra, the monomeric lipid had a typical absorption at 1670 cm^{-1} for the cyclic acrylate carbon-carbon double bond; this absorption totally disappeared as a result of 2 minutes of irradiation, as shown in Figure 3.

The polymerized vesicles showed much greater stability over the unpolymerized ones. On standing at room temperature, the unpolymerized vesicles remained stable only for about 2 days, after which time precipitation of some of the lipid was noted; substantial precipitation of the lipid was observed after 6 days of standing. By comparison, there was only a trace of precipitation in the polymerized vesicle system after standing 6 days at room temperature, and no substantial precipitation was observed after a 2-week period. The electron microscope pictures of Figure 4 showed that most of the polymerized vesicles retained their initial sizes and and shapes after 6 days.

The UV absorption of the aqueous vesicular systems, which provides information on the relative concentration of lipids in aqueous system, also proved the enhanced stability of the polymerized vesicles. The absorptions at 238 nm of the unpolymerized vesicles showed a sharp decrease, as seen in Figure 5, as a result of the precipitation of the lipid in the system. However, the absorptions at the same wavelength of the polymerized system showed a relatively steady trend that meant the polymerized lipid had a longer suspension life in the aqueous system.

The hydrolysis of the connecting groups actually happened both on the monomeric groups and in the polymerized system. Although hydrolysis of the monomeric group in the vesicular system was rapid at high temperature, the rate of hydrolysis at room temperature was quite moderate. The IR spectrum of the recovered monomeric lipid showed that most of the acetal linkages on the monomeric group still survived in an aqueous system after standing in a vesicular system at room temperature for 3 days. However, the IR spectrum of the recovered lipid after 2

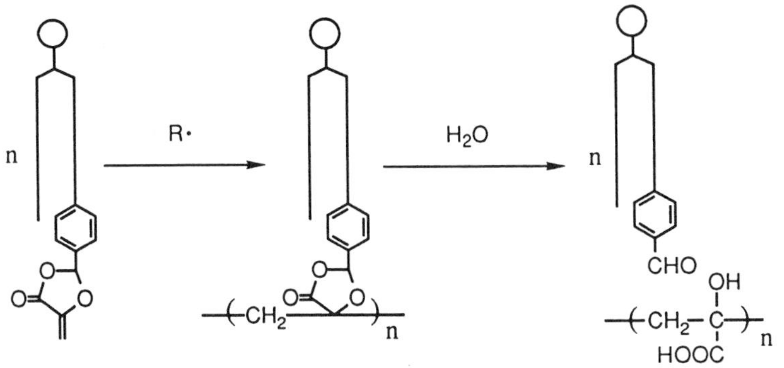

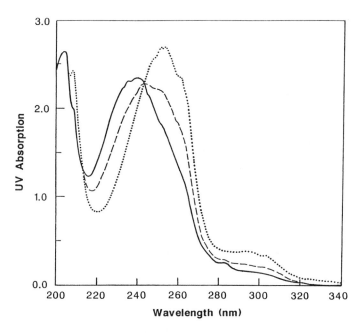

Scheme I

Figure 1. UV spectra of the thermal polymerization system of vesicles of Lipid 1 at 70°C: (-) 0 hr; (---) 3 hr; (••••) 8 hr.

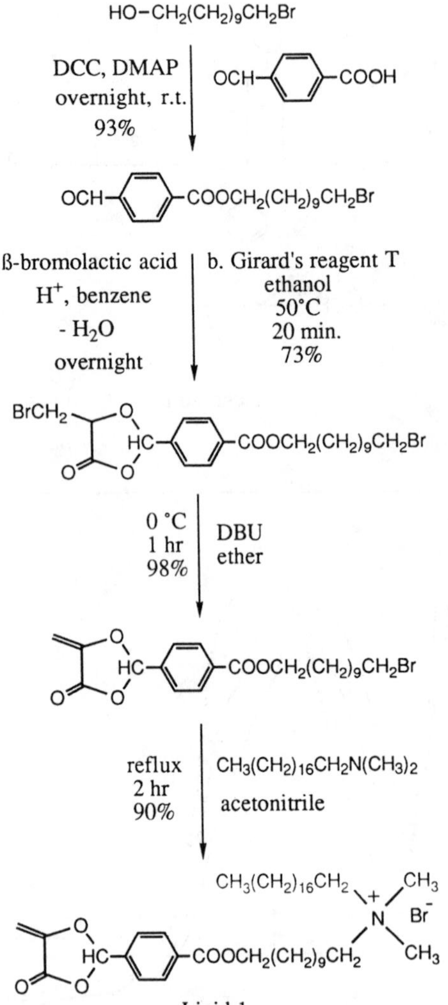

$$HO-CH_2(CH_2)_9CH_2Br$$

DCC, DMAP
overnight, r.t.
93%

$OCH-\!\!\langle\bigcirc\rangle\!\!-COOH$

$OCH-\!\!\langle\bigcirc\rangle\!\!-COOCH_2(CH_2)_9CH_2Br$

a. ß-bromolactic acid | b. Girard's reagent T
H⁺, benzene | ethanol
- H₂O | 50°C
overnight | 20 min.
| 73%

$BrCH_2-\!\!\langle\text{lactide ring}\rangle\!\!-HC-\!\!\langle\bigcirc\rangle\!\!-COOCH_2(CH_2)_9CH_2Br$

0 °C | DBU
1 hr | ether
98% |

$\text{(methylene-dioxolane)}-HC-\!\!\langle\bigcirc\rangle\!\!-COOCH_2(CH_2)_9CH_2Br$

reflux | CH₃(CH₂)₁₆CH₂N(CH₃)₂
2 hr | acetonitrile
90% |

$CH_3(CH_2)_{16}CH_2$
$\text{(methylene-dioxolane)}-HC-\!\!\langle\bigcirc\rangle\!\!-COOCH_2(CH_2)_9CH_2-\overset{+}{N}(CH_3)_2\ Br^-$

Lipid 1

DCC = dicyclohexylcarbodiimide
DMAP = dimethylaminopyridine
DBU = diazabicyclo(5.4.0)undec-7-ene

Scheme II

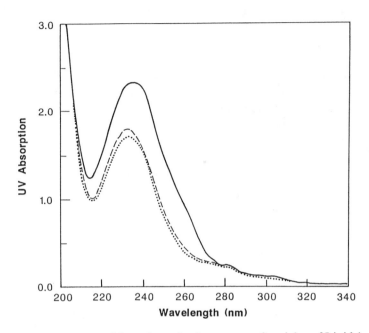

Figure 2. UV spectra of the polymerization system of vesicles of Lipid 1 with UV irradiation: (-) 0 min; (---) 1 min; (••••) 3 min.

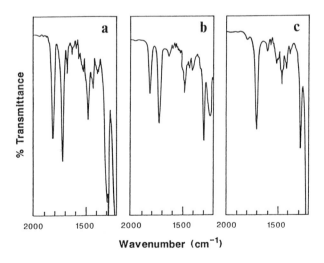

Figure 3. IR spectra of Lipid 1, (a): monomeric lipid, (b): After UV irradiation, (c): hydrolyzed lipid.

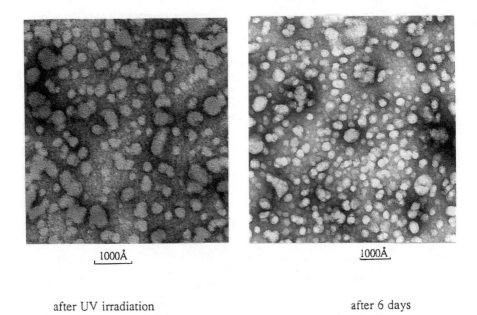

after UV irradiation   after 6 days

Figure 4. Electron micrographs of the polymerized vesicles of Lipid 1 (stained by 1% uranyl acetate).

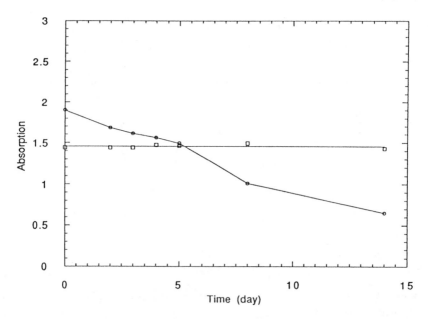

Figure 5. UV absorption at 238 nm of the vesicle systems of Lipid 1 (□: polymerized vesicles;   o: nonpolymerized vesicles).

weeks in an aqueous system showed the hydrolysis of the acetals linkage was almost complete.

The polymerized system gives similar results. The IR spectrum of the polymerized lipid shows two absorption peaks at 1805 cm^{-1} and 1735 cm^{-1} which correspond to the lactone carbonyl and ester groups, respectively. After the polymerized vesicle had been allowed to stand in an aqueous system for 2 weeks, the lactone carbonyl absorption peak at 1805 cm^{-1} disappeared as seen in Figure 3, which indicates the hydrolysis of the connecting acetal linkages has been completed.

In the discussion above it has been shown that the lipid can been polymerized through UV irradiation of its aqueous suspension. The polymerization of the system improves the stability of the synthetic liposomes. Since there is an acetal linkage introduced between the polymer chain and the amphiphilic structure, this linkage can be slowly hydrolyzed in aqueous systems to separate the polymer chain from the lipid.

In the study of polymerizations of the first vesicular system it has been learned that thermal polymerization in an aqueous medium was not possible because of the hydrolysis of the acetal group. UV irradiation seemed the only practical initial way to perform the polymerization. However, it is difficult to follow and manipulate the polymerization, and therefore it is not possible to control the molecular weight of the polymer chain in the vesicular system.

The fluidity is one of the most vital properties of biological membranes. It relates to many functions involved in biological system, and effective biomembrane mimetic chemistry depends on the combination of both stability and mobility of the model membranes. However, in the polymerized vesicles the polymer chain interferes with the motion of the side groups and usually causes a decrease or even the loss of the fluid phases inside the polymerized vesicle (*12, 13*).

In order to retain the fluidity which is a fundamental property of the biological membrane, and to prepare the polymerized vesicles directly from the prepolymerized lipids the incorporation of hydrophilic spacer groups, which decouple the motion of the polymer main chain from the membrane-forming side groups, has been studied (*14, 15*). The advantages of this concept is that loss of mobility and structural changes of the membranes induced by the polymer chain can be avoided, and thus polymerized vesicles can be formed directly from the prepolymerized lipids. Based upon this concept another new polymeric lipid which contains a hydrophilic spacer group derived from ethylene glycol units between the monomeric group of cyclic acrylate and the main amphiphilic structure was synthesized. The synthesis which has been previously published is outlined in Schemes III and IV (*16*).

This polymeric lipid can first be polymerized by free radical initiator in organic solutions before making the vesicles. The proton NMR spectrum of the polymerized lipid shows that vinyl protons of the cyclic acrylate between $\delta 5.00$ ppm and $\delta 6.00$ ppm disappeared from the spectrum, compared with that of monomeric lipid. Also in the IR spectrum (Figure 6) the absorption peak at 1670 cm^{-1} for the cyclic acrylate carbon carbon double bond disappeared as the result of polymerization. The carbonyl absorptions of the esters at 1740 cm^{-1} and the lactone at 1805 cm^{-1} still remain in the spectrum.

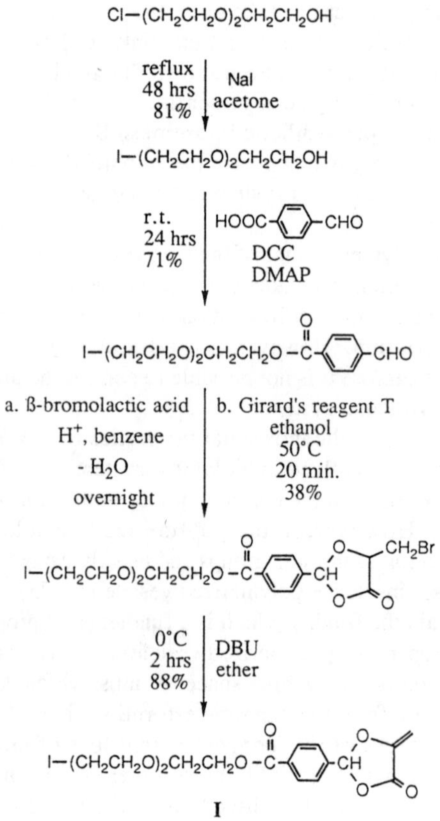

Cl—(CH$_2$CH$_2$O)$_2$CH$_2$CH$_2$OH

reflux | NaI
48 hrs | acetone
81%

I—(CH$_2$CH$_2$O)$_2$CH$_2$CH$_2$OH

r.t. | HOOC—⟨◯⟩—CHO
24 hrs |
71% | DCC
| DMAP

I—(CH$_2$CH$_2$O)$_2$CH$_2$CH$_2$O—C(=O)—⟨◯⟩—CHO

a. ß-bromolactic acid | b. Girard's reagent T
H$^+$, benzene | ethanol
- H$_2$O | 50°C
overnight | 20 min.
| 38%

I—(CH$_2$CH$_2$O)$_2$CH$_2$CH$_2$O—C(=O)—⟨◯⟩—CH⟨ O-CH-CH$_2$Br / O-C=O ⟩

0°C | DBU
2 hrs | ether
88%

I—(CH$_2$CH$_2$O)$_2$CH$_2$CH$_2$O—C(=O)—⟨◯⟩—CH⟨ O-C=CH$_2$ / O-C=O ⟩

I

Scheme III

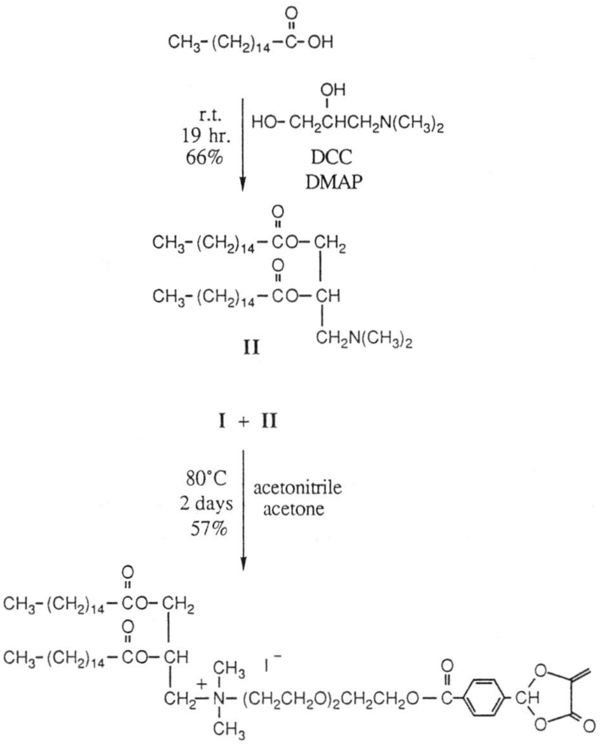

$$CH_3-(CH_2)_{14}-\overset{\overset{\displaystyle O}{\|}}{C}-OH$$

r.t.
19 hr. $HO-CH_2\overset{\overset{\displaystyle OH}{|}}{CH}CH_2N(CH_3)_2$
66%

DCC
DMAP

$$CH_3-(CH_2)_{14}-\overset{\overset{\displaystyle O}{\|}}{C}O-CH_2$$
$$CH_3-(CH_2)_{14}-\overset{\overset{\displaystyle O}{\|}}{C}O-CH$$

II $CH_2N(CH_3)_2$

I + II

80°C   acetonitrile
2 days   acetone
57%

$$CH_3-(CH_2)_{14}-\overset{\overset{\displaystyle O}{\|}}{C}O-CH_2$$
$$CH_3-(CH_2)_{14}-\overset{\overset{\displaystyle O}{\|}}{C}O-CH$$

Lipid 2

Scheme IV

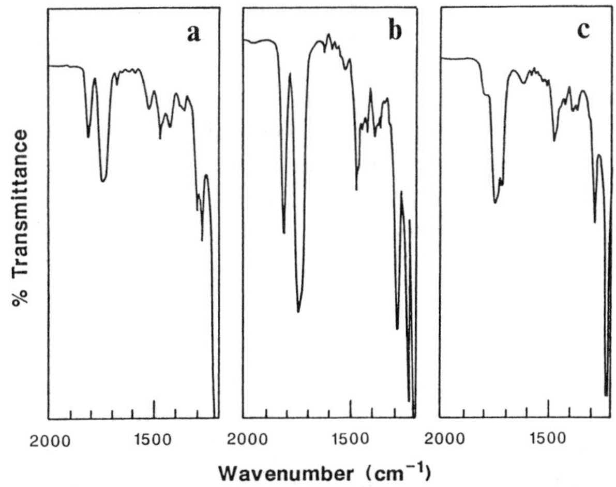

Figure 6. IR spectra of Lipid 2, (a): monomeric lipid, (b): polymerized lipid, (c): hydrolyzed lipid.

The procedure for the formation of vesicles from this prepolymerized lipid was similar to that for the monomeric lipid. However, the concentration of the lipid in this system was lower than in the case of the monomeric lipid. Also the time of sonication for this polymerized lipid was longer than that for the monomeric lipid because of the decreased freedom of motion of the amphiphilic structure in the polymerized system. The electron microscope pictures (Figure 7) show the formation of tiny and very homogeneous vesicles.

The polymerized vesicles from the prepolymerized lipid showed enhanced stability as expected. In a 6-day period there was little precipitate of the lipid. Even after 18 days, as seen in Figure 7, some of the vesicles could be seen in the electron micrograph.

The phase transition of bilayer lipids is related to the highly ordered arrangement of the lipids inside the vesicle. In the ordered gel state below a characteristic temperature, the lipid hydrocarbon chains are in an all-trans configuration. When the temperature is increased, an endothermic phase transition occurs, during which there is a trans-gauche rotational isomerization along the chains which results in a lateral expansion and decrease in thickness of the bilayer. This so-called gel to liquid-crystalline transition has been demonstrated in many different lipid systems and the relationship of the transition to molecular structure and environmental conditions has been studied extensively.

The DSC spectra confirm that the fluid phase of the polymerized vesicles remains and the phase transitions are retained with the introduction of the spacer group. As can been seen in Figure 8 of the DSC spectrum of the monomeric lipid, there is a peak around 28°C which corresponds to the phase transition of monomeric lipid. As the result of the presence of the spacer group, a similar phase transition can also be observed clearly in the spectrum of the polymerized lipid as shown in Figure 9, but the transition temperature is increased to 36°C by the presence of the polymer chains.

The hydrolysis of the cyclic acetal, which was used as the connecting group between the polymer chain and the lipid, was confirmed both by the IR and the proton NMR spectra of the lipid recovered from the vesicular system after standing for 3 weeks at room temperature. The lactone absorption at 1805 cm^{-1} disappeared from the IR spectrum (Figure 6) as the result of hydrolysis. Furthermore, a new aldehyde absorption band at 1705 cm^{-1} was observed in the spectrum, which is related to the substituted benzaldehyde group of the hydrolyzed product. The proton NMR spectrum (Figure 10) also clearly showed the formation of the benzaldehyde, as indicated by the peak at $\delta 10.20$ ppm.

Conclusion

In this paper two new polymerized vesicle systems have been presented. The first lipid can be polymerized in vesicle through UV irradiation. Because the second lipid contains a flexible spacer group it can be prepolymerized in benzene and then converted to vesicles by ultrasonication in water. The polymerization improves the stabilities of the synthetic liposomes. Since there is a acetal linkage between the

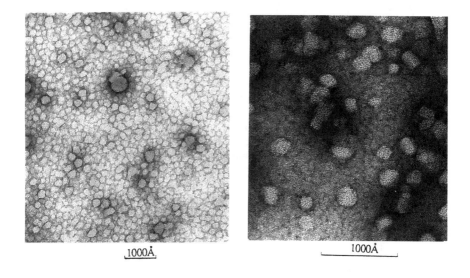

polymerized vesicles   after 18 days

Figure 7. Electron micrographs of the vesicles of the polymerized Lipid 2 (stained by 1% uranyl acetate).

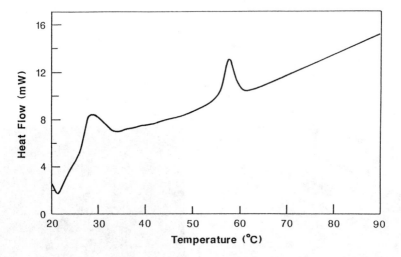

Figure 8. DSC spectrum of the monomeric Lipid 2 in an aqueous system.

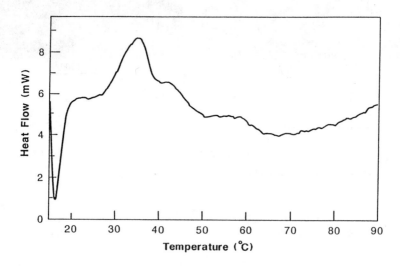

Figure 9. DSC spectrum of the polymerized Lipid 2 in an aqueous system.

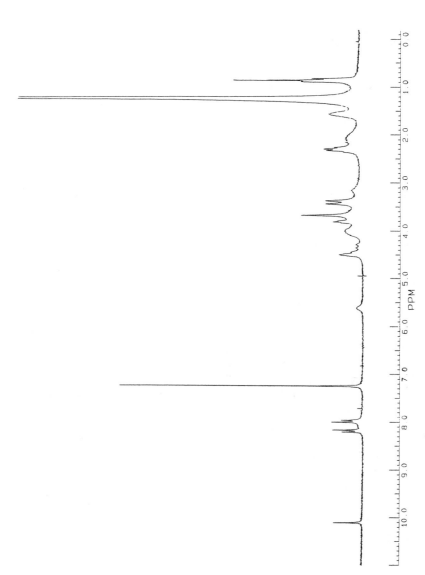

Figure 10. ^{1}H NMR spectrum of the recovered polymerized Lipid 2 in a vesicular system after 3 weeks.

polymer chain and the amphiphilic structure, this linkage can be slowly hydrolyzed in an aqueous system to separate the polymer chain from the lipid and finally generate a water-soluble biodegradable polymer.

Acknowledgments

We are grateful to the Office of Naval Research, the Polymer Program of the National Science Foundation and Interx Research Corporation for their partial support of this research. We are also grateful to the Naval Research Laboratory and EM Laboratory of the Department of Zoology at the University of Maryland at College Park for their cooperation in part of this research. We also would like to thank Dr. Jacques Roovers for reviewing this paper.

Literature Cited

1. Gros, L.; Ringsdorf, H.; Schupp, H. *Angew Chem., Int. Ed. Engl.* **1981**, *20*, 305.
2. Regen, S. L.; Czech, B.; Singh, A. *J. Am. Chem. Soc.* **1980**, *102*, 6638.
3. Akimoto, A.; Norn, K.; Gros, L.; Ringsdorf, H.; Schupp, H. *Angew Chem. Int. Ed. Engl.* **1981**, *20*, 90.
4. Tundo, P.; Kippenberger, D. J.; Klahn, P. L.; Pietro, N. E.; Tao, T. C.; Fendler, J. H. *J. Am. Chem. Soc.* **1982**, *104*, 456.
5. Samuel, N. K. P.; Singh, M.; Yamagachi, K.; Regen, S. L. *J. Am. Chem. Soc.* **1985**, *107*, 42.
6. Weber, B. A.; Dodrer, N.; Regen, S. L. *J. Am. Chem. Soc.* **1987**, *109*, 4419.
7. Neumann, R.; Ringsdorf, H. *J. Am. Chem. Soc.* **1986**, *108*, 4487.
8. Bailey, W. J.; Chou, J. L.; Feng, P-.Z.; Kuruganti, V.; Zhou, L. -L. *Acta Polym.* **1988**, *39*, 335.
9. Hoechst, A. G. U.S. Patent 3,890,288, **1975**.
10. Mulders, J.; Gilain, J. *J. Water Res.* **1977**, *11*, 571.
11. Bailey, W. J.; Zhou, L.-L. *Polym. Prep.,* Am. Chem. Soc., Div. Polym. Chem. **1988**, *29(2)*, 300.
12. Ruschl, R.; Folda, T.; Ringsdorf, H. *Makromol. Chem. Suppl.* **1984**, *6*, 245.
13. Bader, H.; Dorn, K.; Hupfer, B.; Ringsdorf, H. *Adv. Polym. Sci.* **1985**, *64*, 1.
14. Elbert, R.; Laschewsky, A.; Ringsdorf, H. *J. Am. Chem. Soc.* **1985**, *107*, 4134.
15. Laschewsky, A.; Ringsdorf, H.; Schmidt, G.; Schneider, J. *J. Am. Chem. Soc.* **1987**, *109*, 788.
16. Bailey, W. J.; Zhou, L.-L. *Polym. Prep.,* Am. Chem. Soc.., Div. Polym. Chem. **1988**, *29(2)*, 302.

RECEIVED March 19, 1991

INDEXES

Author Index

Affiliation Index

Subject Index

Production: Margaret J. Brown and Donna Lucas
Indexing: Deborah H. Steiner
Acquisition: A. Maureen Rouhi
Cover design: Peggy Corrigan

Printed and bound by Maple Press, York, PA

Highlights from ACS Books

Good Laboratory Practices: An Agrochemical Perspective
Edited by Willa Y. Garner and Maureen S. Barge
ACS Symposium Series No. 369; 168 pp; clothbound, ISBN 0–8412–1480–8

Silent Spring Revisited
Edited by Gino J. Marco, Robert M. Hollingworth, and William Durham
214 pp; clothbound, ISBN 0–8412–0980–4; paperback, ISBN 0–8412–0981–2

Insecticides of Plant Origin
Edited by J. T. Arnason, B. J. R. Philogène, and Peter Morand
ACS Symposium Series No. 387; 214 pp; clothbound, ISBN 0–8412–1569–3

Chemistry and Crime: From Sherlock Holmes to Today's Courtroom
Edited by Samuel M. Gerber
135 pp; clothbound, ISBN 0–8412–0784–4; paperback, ISBN 0–8412–0785–2

Handbook of Chemical Property Estimation Methods
By Warren J. Lyman, William F. Reehl, and David H. Rosenblatt
960 pp; clothbound, ISBN 0–8412–1761–0

The Beilstein Online Database: Implementation, Content, and Retrieval
Edited by Stephen R. Heller
ACS Symposium Series No. 436; 168 pp; clothbound, ISBN 0–8412–1862–5

Materials for Nonlinear Optics: Chemical Perspectives
Edited by Seth R. Marder, John E. Sohn, and Galen D. Stucky
ACS Symposium Series No. 455; 750 pp; clothbound; ISBN 0–8412–1939–7

Polymer Characterization:
Physical Property, Spectroscopic, and Chromatographic Methods
Edited by Clara D. Craver and Theodore Provder
Advances in Chemistry No. 227; 512 pp; clothbound, ISBN 0–8412–1651–7

From Caveman to Chemist: Circumstances and Achievements
By Hugh W. Salzberg
300 pp; clothbound, ISBN 0–8412–1786–6; paperback, ISBN 0–8412–1787–4

The Green Flame: Surviving Government Secrecy
By Andrew Dequasie
300 pp; clothbound, ISBN 0–8412–1857–9

For further information and a free catalog of ACS books, contact:
American Chemical Society
Distribution Office, Department 225
1155 16th Street, NW, Washington, DC 20036
Telephone 800–227–5558

Bestsellers from ACS Books

The ACS Style Guide: A Manual for Authors and Editors
Edited by Janet S. Dodd
264 pp; clothbound, ISBN 0–8412–0917–0; paperback, ISBN 0–8412–0943–X

Chemical Activities and Chemical Activities: Teacher Edition
By Christie L. Borgford and Lee R. Summerlin
330 pp; spiralbound, ISBN 0–8412–1417–4; teacher ed. ISBN 0–8412–1416–6

Chemical Demonstrations: A Sourcebook for Teachers,
Volumes 1 and 2, Second Edition
Volume 1 by Lee R. Summerlin and James L. Ealy, Jr.;
Vol. 1, 198 pp; spiralbound, ISBN 0–8412–1481–6;
Volume 2 by Lee R. Summerlin, Christie L. Borgford, and Julie B. Ealy
Vol. 2, 234 pp; spiralbound, ISBN 0–8412–1535–9

Writing the Laboratory Notebook
By Howard M. Kanare
145 pp; clothbound, ISBN 0–8412–0906–5; paperback, ISBN 0–8412–0933–2

Developing a Chemical Hygiene Plan
By Jay A. Young, Warren K. Kingsley, and George H. Wahl, Jr.
paperback, ISBN 0–8412–1876–5

Introduction to Microwave Sample Preparation: Theory and Practice
Edited by H. M. Kingston and Lois B. Jassie
263 pp; clothbound, ISBN 0–8412–1450–6

Principles of Environmental Sampling
Edited by Lawrence H. Keith
ACS Professional Reference Book; 458 pp;
clothbound; ISBN 0–8412–1173–6; paperback, ISBN 0–8412–1437–9

Biotechnology and Materials Science: Chemistry for the Future
Edited by Mary L. Good (Jacqueline K. Barton, Associate Editor)
135 pp; clothbound, ISBN 0–8412–1472–7; paperback, ISBN 0–8412–1473–5

Personal Computers for Scientists: A Byte at a Time
By Glenn I. Ouchi
276 pp; clothbound, ISBN 0–8412–1000–4; paperback, ISBN 0–8412–1001–2

Polymers in Aqueous Media: Performance Through Association
Edited by J. Edward Glass
Advances in Chemistry Series 223; 575 pp;
clothbound, ISBN 0–8412–1548–0

For further information and a free catalog of ACS books, contact:
American Chemical Society
Distribution Office, Department 225
1155 16th Street, NW, Washington, DC 20036
Telephone 800–227–5558